Polarization Optics of Random Media

Springer
Berlin
Heidelberg
New York
Hong Kong
London
Milan
Paris
Tokyo

Alexander A. Kokhanovsky

Polarization Optics of Random Media

Springer

Published in association with

Chichester, UK

Dr Alexander A. Kokhanovsky
B. I. Stepanov Institute of Physics   and Institute of Remote Sensing
National Academy of Sciences of Belarus   Bremen University
Minsk   Bremen
Belarus   Germany

SPRINGER–PRAXIS BOOKS IN ENVIRONMENTAL SCIENCES
SUBJECT *ADVISORY EDITOR*: John Mason M.Sc., B.Sc., Ph.D.

ISBN 3-540-42635-3 Springer-Verlag Berlin Heidelberg New York

Bibliographic information published by Die Deutsche Bibliothek

Die Deutsche Bibliothek lists this publication in the Deutsche Nationalbibliografie; detailed bibliographic data are available from the Internet at http://dnb.ddb.de

Library of Congress Cataloging-in-Publication Data
Kokhanovsky, Alexander A.
  Polarization optics of random media / Alexander A. Kokhanovsky.
     p. cm. – (Springer-Praxis books in environmental sciences)
  Includes bibliographical references and index.
  ISBN 3-540-42635-3 (acid-free paper)
  1. Polarization (Light)   2. Polarimetry.   3. Particles–Optical.
  4. Light–Scattering   5. Radiative transfer.   I. Title.   II. Series.

QC441.K65 2003
535.5$'$–dc21
                                                    2003050353

Apart from any fair dealing for the purposes of research or private study, or criticism or review, as permitted under the Copyright, Designs and Patents Act 1988, this publication may only be reproduced, stored or transmitted, in any form or by any means, with the prior permission in writing of the publishers, or in the case of reprographic reproduction in accordance with the terms of licences issued by the Copyright Licensing Agency. Enquiries concerning reproduction outside those terms should be sent to the publishers.

© Praxis Publishing Ltd, Chichester, UK, 2003
Printed in Germany

The use of general descriptive names, registered names, trademarks, etc. in this publication does not imply, even in the absence of a specific statement, that such names are exempt from the relevant protective laws and regulations and therefore free for general use.

Cover design: Jim Wilkie
Project Management: Originator Publishing Services, Gt Yarmouth, Norfolk, UK

Printed on acid-free paper

Contents

Preface ix

List of figures xi

List of tables xv

1 **Introduction** 1
   1.1 Discrete random media 1
   1.2 Light beams 5
   1.3 The interaction matrix 10
   1.4 Further reading 12

2 **Polarized radiative transfer** 13
   2.1 Vector radiative transfer equation 13
   2.2 Direct light 18
   2.3 Diffused light 20
        2.3.1 Thin layers 20
        2.3.2 Thick layers 25
   2.4 Further reading 33

3 **Local optical characteristics** 37
   3.1 General equations 37
        3.1.1 The phase matrix 37
        3.1.2 The extinction matrix 42
        3.1.3 The classification of disperse media 45
   3.2 Spherical particles 47
        3.2.1 Mie theory 47

		3.2.2	Optical properties of spherical polydispersions	48
		3.2.3	Bubbles	75
	3.3	Ellipsoidal particles		86
		3.3.1	General remarks	86
		3.3.2	Rayleigh ellipsoids	87
		3.3.3	Spheroidal particles	93
	3.4	Cylinders		104
	3.5	Irregularly shaped particles		110
		3.5.1	Koch fractals	110
		3.5.2	Experimental results	118
		3.5.3	Gaussian random particles	123
	3.6	Inhomogeneous particles		126
		3.6.1	General equations	126
		3.6.2	Rayleigh approximation	127
		3.6.3	Rayleigh–Gans approximation	128
		3.6.4	Polydispersions of coated spheres	129
	3.7	Optically active particles		132
		3.7.1	General equations	132
		3.7.2	The extinction matrix	136
		3.7.3	Rayleigh approximation	146
	3.8	Further reading		148
4	**Environmental polarimetry**			**153**
	4.1	Clouds		153
		4.1.1	Water clouds	153
		4.1.2	Ice clouds	157
	4.2	Aerosols		159
	4.3	Natural surfaces		170
		4.3.1	Land	171
		4.3.2	Ocean	174
		4.3.3	Snow and ice	179
	4.4	Polarimetric remote sensing		180
	4.5	Chiro-optical spectroscopy (CS) of turbid media		184
		4.5.1	Introduction	184
		4.5.2	CD and ORD spectra for turbid layers with non-spherical particles	186
		4.5.3	Cylinders	188
		4.5.4	Ellipsoids	189
	4.6	Further reading		193

Appendix 1 The tensor radiative transfer equation ... 197

Appendix 2 Jones matrices, Mueller matrices, and Stokes vectors ... 203

Appendix 3 The system of linear differential equations. 207

Appendix 4 Local optical characteristics of cloudy media 211

Appendix 5 Fresnel equations 213

Index 217

Preface

A light beam is characterized by its propagation direction, frequency range, intensity, and polarization. The subject of this book is light polarization and its change due to the interaction of light beams with discrete random media (e.g., clouds and fogs).

The polarization of light was discovered by Erasmus Bortolinus in 1669. This phenomenon can be easily demonstrated using a sheet of Polaroid. In particular, putting the polarizer perpendicular to a light beam and rotating it, allows observation of a number of interesting phenomena in the transmitted light. For instance, incident light beams having the same frequency and intensity can in principle produce different output beams. For some types of incident beams and some orientations of a polarizer, the transmitted light beam can almost totally disappear. In other cases the rotation of a polarizer may not influence the output beam intensity at all. This suggests that a full characterization of a light beam should include the description of its properties in the plane perpendicular to the propagation direction.

Visible light is a particular case of an electromagnetic wave. These waves are transverse, which means that oscillations occur in the plane perpendicular to the direction of propagation. This is similar to the case of waves on the surface of a calm lake disturbed by rain droplets. These waves propagate to the shores in the horizontal plane, but the oscillations occur in the vertical plane.

The subject of this book is the study of the changes in light polarization due to the interaction of light beams with random discrete media such as oceanic suspensions, cloud and snow fields, aerosols, and bio-liquids. Different media change not only the intensity but also the polarization of incident waves in different ways. This is used as an important tool for technological and remote sensing applications. For instance, very thin layers of ice crystals in the upper troposphere can not be detected by lidar stations if the receiver detects only the light polarized in the same plane as an initial laser beam. Cross-polarization measurements, however, reveal their existence.

Moreover, this also gives us information on the shape of particles. There is no cross polarization of light in the exact backscattering direction for spherical particles.

Initially, unpolarized solar incident light becomes polarized due to molecular and particle single light scattering in the terrestrial atmosphere. This polarization may be diminished due to multiple light scattering. Both single and multiple light scattering processes are considered in this book.

In Chapter 1 we introduce the basic notions of the theory, like particle size distributions, Stokes parameters, and the interaction matrix. Chapter 2 gives the main equations that govern polarized light transfer in random media. Limiting cases of thin and thick media are considered in detail. Chapter 3 is devoted to a study of polarization characteristics of singly scattered light as functions of the size, shape, structure, and electric properties (e.g., refractive indices) of particles. General results presented in Chapters 2 and 3 are used to describe the polarization characteristics of various natural media (e.g., clouds and aerosols) in Chapter 4.

This work has benefited from discussions with a number of people. It is not possible to mention all of them here, but I am deeply grateful for their helpful hints and comments. The use of codes, developed by Andreas Macke (single light scattering) and Kazuhiko Masuda (multiple light scattering) is acknowledged with many thanks. The author is thankful to Eleonora Zege for originating his interest in the polarization optics of random discrete media.

This work was conducted while the author held fellowships at various scientific centres. The author is grateful to John P. Burrows (University of Bremen, Germany), Arkadii P. Ivanov (Institute of Physics, Belarus), Alan R. Jones (Imperial College, England), Teruyuki Nakajima (University of Tokyo, Japan), and Reiner Weichert (Clausthal Technical University, Germany) for their cooperation and support during the author's work at their institutions.

Figures

1.1	The dependence of the azimuth ψ on the ratio U/Q.	8
1.2	The dependence of the ellipticity angle φ on the ratio V/w	9
1.3	The dependence of the ellipticity e on the ellipticity angle φ	10
2.1	The geometry of the problem	16
2.2	The dependence of the degree of polarization of reflected light on the incidence angle at the nadir observation for the particle size distribution (PSD)	31
2.3	The same as in Figure 2.2 but as the function of the inverse optical thickness at the incidence angle $37°$.	32
3.1	The dependence of the phase function of spherical polydispersions, the degree of polarization p_l, p_{44}, p_{34}, q, and s with the PSD (1.5) at $\mu = 6$ on the effective radius of droplets $a_{ef} = 0.06$ μm, 0.6 μm, 6 μm, and 15 μm	51
3.2	The dependence of the phase function of spherical polydispersions, the degree of polarization p_l, p_{44}, and p_{34} with the PSD (1.5) at $a_{ef} = 6$ μm and $\mu = 6$ and 12	54
3.3	The dependence of the phase function of spherical polydispersions, the degree of polarization p_l, p_{44}, and p_{34} with the PSD (1.5) at $a_{ef} = 6$ μm	57
3.4	The dependence of the phase function of spherical polydispersions, the degree of polarization p_l, p_{44}, p_{34}, q, and s with the PSD (1.5) at $a_{ef} = 6$ μm and $\mu = 6$ on the real part of the refractive index $n = 1.1$, 1.333, 1.53 and 1.7	59
3.5	The primary and secondary rainbow angles as functions of the refractive index n and the dependence of the separation between primary and secondary rainbows $\Delta\theta$ on the refractive index n.	70
3.6	The phase function of spherical polydispersions, the degree of polarization p_l, p_{44}, and p_{34} with the PSD (1.5) at $a_{ef} = 15$ μm, and $\mu = 6$, calculated with the Mie theory and the geometrical optics approach at $\lambda = 0.55$ μm and $n = 1.333$	71
3.7	The phase function of spherical polydispersions, the degree of polarization p_l, p_{44}, and p_{34} with the PSD (1.5) at $a_{ef} = 15$ μm and $\mu = 6$, calculated with the Mie theory and the geometrical optics approach at $\lambda = 0.55$ μm and $n = 1.52$	73

xii Figures

3.8	The phase function of spherical polydispersions of bubbles and droplets, the degree of polarization p_l, p_{44}, p_{34}, q, and s with the PSD (1.5) at $a_{ef} = 15\,\mu\text{m}$ and $\mu = 6$, obtained with the Mie theory	76
3.9	The phase function of spherical polydispersions of bubbles and droplets, p_{12}, p_{44}, and p_{34} with the PSD (1.5) at $a_{ef} = 15\,\mu\text{m}$ and $a_{ef} = 30\,\mu\text{m}$, obtained with the Mie theory	80
3.10	The phase function, p_{12}, p_{44}, and p_{34} of spherical non-absorbing bubbles obtained in the framework of the geometrical optics approximation at the refractive index equal to 0.3, 0.4, 0.5, 0.6, 0.75 (air bubbles in water), 0.8, and 0.9	82
3.11	The dependence of the asymmetry parameter g on the refractive index of bubbles n	85
3.12	The phase function and degree of polarization for $y = 1, 2, 4, 13$	89
3.13	The phase function $p(\theta) = a_1$, the matrix elements $p_{22} = a_2/a_1$, $p_{33} = a_3/a_1$, $p_{44} = a_4/a_1$, $-p_{12} = -b_1/a_1$, and $p_{34} = b_2/a_1$ versus scattering angle θ for spheres and surface-equivalent randomly oriented spheroids	96
3.14	The dependence of the phase function $p(\theta)$, p_{12}, p_{22}, p_{33}, p_{44}, and p_{34} on the axis ratio $\xi = 0.05, 0.5, 1.0, 2.0$, and 20.0 for non-absorbing spheroidal particles at $n = 1.5$, obtained using geometrical optics calculations	98
3.15	The phase function $p(\theta) = F_{11}$, normalized matrix elements $p_{32} = F_{33}/F_{11}$, $-p_{12} = -F_{12}/F_{11}$, $p_{22} = F_{22}/F_{11}$, $p_{44} = F_{44}/F_{11}$, and $p_{34} = -F_{34}/F_{11}$ of randomly oriented oblate spheroids and spheres.	102
3.16	The phase function $p(\theta) = F_{11}$, normalized matrix elements $p_{33} = F_{33}/F_{11}$, $-p_{12} = -F_{12}/F_{11}$, $p_{22} = F_{22}/F_{11}$, $p_{44} = F_{44}/F_{11}$, and $p_{34} = -F_{34}/F_{11}$ of spheres and volume-equivalent randomly oriented oblate spheroids	103
3.17	The phase function $p(\theta) = F_{11}$, normalized matrix elements $p_{33} = F_{33}/F_{11}$, $-p_{12} = -F_{12}/F_{11}$, $p_{22} = F_{22}/F_{11}$, $p_{44} = F_{44}/F_{11}$, and $p_{34} = -F_{34}/F_{11}$ of randomly oriented infinite cylinders and randomly oriented prolate spheroids.	105
3.18	The phase function $p(\theta) = F_{11}$, normalized matrix elements $p_{33} = F_{33}/F_{11}$, $-p_{12} = -F_{12}/F_{11}$, $p_{22} = F_{22}/F_{11}$, $p_{44} = F_{44}/F_{11}$, and $p_{34} = F_{34}/F_{11}$ of randomly oriented infinite cylinders and spheres.	106
3.19	The same as in Figure 3.16 but for randomly oriented circular cylinders	108
3.20	The phase function of randomly oriented non-absorbing finite hexagonal cylinders.	109
3.21	The same as in Figure 3.20 but for the normalized phase matrix elements . . .	109
3.22	The phase function, p_{12}, p_{22}, p_{33}, p_{44}, and p_{34} of fractals for various values of the real part of the refractive index.	112
3.23	The dependence of the parameters q and r on the scattering angle for ice clouds and dust aerosols	117
3.24	The comparison of calculations and measurements of the normalized phase matrix elements.	119
3.25	Normalized matrix elements, obtained from measurements and calculations at $n = 1.1$	120
3.26	Normalized matrix elements, obtained from measurements and calculations at the complex refractive index $m = 1.4$.	121
3.27	The same as in Figure 3.26, but for the averaged normalized phase matrix of aerosol particles	122
3.28	The phase function, normalized at the scattering angle of $30°$, obtained from measurements and calculated using the fractal particle model	123

Figures xiii

3.29	The comparison of fractal and Gaussian particle models for the calculation of the phase function.	125
3.30	The light scattered intensity I for spherical polydispersions of coated spheres	130
3.31	The angular dependence of the phase function, the normalized phase matrix elements, and the optical rotatory dispersion (ORD) and circular dichroism (CD) spectra of PGA spheres.	137
3.32	The spectral dependence of the average real and imaginary parts of the refractive index of the PGA.	143
3.33	The ORD and CD spectra of the PGA in solution and particulate media made of the PGA spheres and the ORD and CD spectrums for the PGA in solution and PGA spheres.	145
4.1	The dependence of the intensity and the degree of polarization of reflected light on the zenith observation angle with the sun overhead.	154
4.2	The same as in Figure 4.1 but for the fixed cloud optical thickness equal to 32 and different effective radii.	155
4.3	The same as Figure 4.1 with a different wavelength ($\lambda = 2.25\,\mu m$).	156
4.4	The same as Figure 4.1 with a different wavelength ($\lambda = 3.4\,\mu m$).	156
4.5	Phase function and degree of linear polarization patterns for a typical cirrostratus.	157
4.6	The upper panel shows linear polarization of sunlight reflected from cirrus clouds and the lower panels display full polarization observed from the polarimeter POLDER.	158
4.7	The phase function, p_l, p_{44}, p_{34}, q, and s of the water-soluble, soot, oceanic, and dust aerosols at the wavelength 0.55 μm.	160
4.8	The phase function, the degree of polarization p_l, p_{44}, p_{34}, q, and s of the water-soluble, soot, oceanic, and dust aerosols at the wavelength 2.25 μm.	163
4.9	The dependence $q(\theta)$ according to measurements for different weather conditions at the wavelength 0.55 μm.	168
4.10	Distribution of the degree of polarization along the sun's vertical axis in a Rayleigh atmosphere and the degree of linear polarization of zenith skylight.	169
4.11	Typical reflectance spectra of land and water surfaces.	172
4.12	Degree of polarization as a function of angle in the principle plane of light reflected from various types of mineral surfaces.	172
4.13	Degree of polarization as a function of the angle in the principal plane for light of five different wavelengths reflected from the Mojave Desert sand.	173
4.14	Degree of polarization as a function of angle in the principal plane of light of four different wavelengths reflected from black loam soil.	173
4.15	The Voss and Fry (1984) data for the degree of polarization, p_{22}, p_{33}, of light scattered by an elementary volume of oceanic water.	177
4.16	The bidirectional polarized reflectance as the function of the scattering angle for semi-infinite snow surface.	179
4.17	The error of retrieval of values of effective radius and cloud optical thickness as a function of the measurement error.	183
4.18	The dependence of the reflection function on the degree of polarization at the rainbow scattering angle for various cloud optical thickness and effective radii of droplets.	184
4.19	The dependence of u on ρ.	191
4.20	The dependence of v on ρ.	191

4.21	The ORD spectrum for various values of $b = 2\pi d(\bar{m} - 1)$.	192
4.22	The same as in Figure 4.21 but for the CD spectrum	192
A5.1	The dependence of the degree of polarization of the reflected light on the incidence angle for selected values of the refractive index	215
A5.2	The same as in Figure A5.1 but for the transmitted light	215

Tables

1.1	The expressions for volumes, surface areas, and effective radii of particles having different shapes.	5
2.1	The dependence K_{01} on the cosine of the observation angle μ.	29
2.2	The dependence K_{02} on the cosine of the observation angle μ.	29
2.3	The degree of linear polarization obtained from data given in Tables 2.1 and 2.2	29
2.4	The dependence of functions $R^0_{\infty 11}(1,\mu)$, $R^0_{\infty 21}(1,\mu)$, and $p_l(1,\mu)$ at the nadir observation on the cosine of the observation angle μ for the model of venerian particles.	30
2.5	The dependence of functions $R^0_{\infty 11}(1,\mu)$, $R^0_{\infty 21}(1,\mu)$, and $p_l(1,\mu)$ at the nadir observation on the cosine of the observation angle μ for Rayleigh atmosphere	30
3.1	General forms of the normalized phase matrices for isotropic symmetrical and asymmetrical media.	46
3.2	As Table 3.1 but for spherical particles.	46
3.3	Light scattering characteristics for isotropic homogeneous spheres.	48
3.4	The dependence of θ_p, θ_s, and $\Delta\theta$ on n.	69
3.5	The asymmetry parameter g for various n smaller than 1	84
3.6	The values of L_j for different shapes of particles	93
3.7	The dependence of the halo angle for the hexagonal cylinder of the real part of the refractive index n.	110
3.8	The asymmetry parameters g_0 and g for various n.	115
3.9	Parameters of the refractive index parameterization for the case of PGA.	143
4.1	Parameters of particle size distributions of selected aerosol types.	159
4.2	Aerosol models.	166
A2.1	Jones and Mueller matrices	203
A2.2	Stokes vectors.	205
A4.1	Parameters p_n at the wavelength $1.55\,\mu m$.	211
A4.2	Parameters q_n for different wavelengths λ	211

Have not the Rays of Light several sides, endued with several original properties?

Issac Newton

1

Introduction

1.1 DISCRETE RANDOM MEDIA

Most of the natural media surrounding us in daily life are characterized by a random distribution of their structural elements in space and time domains. They are usually called random media. Let us take a cloud in the sky. It consists of collections of water droplets, ice crystals, or both. The positions of the cloud particles are not correlated in space. This is in sharp contrast to the case of an ideal crystal where all the atoms are situated at given prescribed places, forming a so-called crystalline lattice. The internal geometry of the medium dictates (among other factors) its physical and, in particular, optical properties. Therefore, random media of a different nature share a lot of common properties. In particular, the theoretical tools to study them are largely based on the statistical approach. The notions of probability and statistical distribution lie in the very heart of random media physics.

The subject of this book is the optics of random media. As a matter of convenience these are divided into two broad classes, namely continuous and discrete random media. As far as optics is concerned, they differ by the behaviour of the refractive index m. For continuous media, m is a continuous (but random) function of the space and time coordinates. This is not the case for discrete random media, where m does not change continuously. A water cloud is an example of a discrete random medium. The refractive index changes abruptly from that of air to that of water throughout the cloud. Conversely, distillate water in a cuvette is an example of a continuous random medium. Here foreign particles are not present and randomness is entirely due to fluctuations of temperature and pressure. These fluctuations are distributed randomly and so cause random fluctuations of the refractive index.

This division of media into two classes is convenient for theoretical studies and some particular applications. In reality, however, we often encounter the case when discrete particles are randomly distributed inside a continuous random medium. Examples are the terrestrial atmosphere and ocean.

Depending on the problem, one can apply continuum or random media optical techniques. For instance, in studies of light reflection from optically thick clouds, one can neglect fluctuations in the surrounding air. In particular, the contribution of fluctuations is weak in comparison with multiple light scattering by large droplets. On the other hand, if we are interested in the characterization of the clear air turbulence, we can ignore discrete particles in a first approximation.

We stress, however, that while analysing experimental data the model of discrete particles embedded in a continuous random medium should be used. This allows for a more precise description of the phenomena studied. The subject of this book is the optics of discrete random media. In physical chemistry, such media are also called disperse media.

Disperse media are defined as collections of small particles in a host medium. Examples are clouds and aerosols, snow, oceanic water, foam, ice, and interstellar dust, to name a few. Particles can have different size, shape, orientation, internal structure, and chemical composition. Oriented systems of non-spherical particles belong to the class of so-called anisotropic media. Their optical properties depend on the direction of light propagation.

In most cases, a disperse medium is a mixture of particles with different chemical compositions (e.g., dust aerosols). Particles in a mixture can have different sizes, shapes, and orientations, depending on their origin and history. For example, one can find soot and dust particles, sulphate and organic aerosols, bio-particles and volcanic ash in the Earth's atmosphere. Liquid particles are mostly of a spherical shape, but this is not the case for solid particles, like ice crystals, dust grains, and snow flakes.

The prime aim of this book is to study the response of disperse systems to incident electromagnetic waves (mostly in the optical band of the electromagnetic spectrum). This response depends on the number concentration, the refractive index, size, shape, internal structure, and orientation of particles. The number concentration is defined as the number of particles in a unit volume, and can vary from a few particles to hundred thousands (per cm^3) depending on their size, origin, and many other factors. The refractive indices (or correspondent tensors for anisotropic and gyrotropic particles) $m = n - i\chi$, are determined by the chemical composition of particles and the wavelength λ of the incident radiation. For instance, refractive indices of different substances have approximately the following ranges at the spectral interval 0.4–0.8 µm: $n \approx 1.34$–1.33, $\chi = 1.86 \times 10^{-9} - 1.25 \times 10^{-7}$ (for water); $n \approx 1.32$–$1.31, \chi = 3.0 \times 10^{-9} - 1.34 \times 10^{-7}$ (for ice); $n \approx 1.53, \chi = 0.008$ (for dust); and $n \approx 1.75$, $\chi = 0.46 \div 0.43$ (for soot).

The distribution of particle sizes, shapes, and orientations can be described by the distribution function $f(\vec{a})$ with the normalization condition:

$$\int f(\vec{a})\, d\vec{a} = 1, \qquad (1.1)$$

where the vector-parameter $\vec{a}$ (with components a_j) is determined by the size, shape, and orientation of particles. For spherical particles, the vector-parameter $\vec{a}$ reduces to the scalar value a (the radius of particles). The function $f(a)$ is called the particle size distribution (PSD) in this specific case. It can be uni- or multi-modal. The most important parameters of the particle size distributions are the mean radius

$$\bar{a} = \int_0^\infty a f(a)\, da, \tag{1.2}$$

the coefficient of variance

$$\Delta = \frac{\sqrt{\overline{(a-\bar{a})^2}}}{\bar{a}}, \tag{1.3}$$

and the effective radius

$$a_{ef} = \frac{\overline{a^3}}{\overline{a^2}}, \tag{1.4}$$

where $\overline{a^2} = \int_0^\infty a^2 f(a)\, da$, $\overline{a^3} = \int_0^\infty a^3 f(a)\, da$, and $\overline{(a-\bar{a})^2} = \int_0^\infty (a-\bar{a})^2 f(a)\, da$.

In many cases the optical characteristics of disperse systems are not sensitive to the fine structure of the PSDs and are determined mostly by the ratio (ν) of the average volume of the particles to the average surface area ($\nu = a_{ef}/3$ for spheres) and the coefficient of variance Δ.

The most often used forms of particle size distributions are described by the following equations:

$$f(a) = \frac{\mu^{\mu+1}}{a_0^{\mu+1}\Gamma(\mu+1)} a^\mu e^{-\mu(a/a_0)} \tag{1.5}$$

and

$$f(a) = \frac{1}{\sqrt{2\pi}a\sigma} \exp\left[-\frac{\ln^2(a/a_m)}{2\sigma^2}\right], \tag{1.6}$$

where a_0, a_m, μ, and σ are parameters of PSDs and $\Gamma(\mu+1)$ is the Gamma function. Equation 1.5 is called the gamma distribution and Equation 1.6 is the log-normal distribution. The values of a_{ef} and Δ for both distributions are given by the following equations:

$$a_{ef} = a_0\left(1 + \frac{3}{\mu}\right), \qquad \Delta = \frac{1}{\sqrt{1+\mu}}$$

for the gamma PSD and

$$a_{ef} = a_m \exp\left(\frac{5\sigma^2}{2}\right), \qquad \Delta = \sqrt{\exp(\sigma^2) - 1}$$

for the log-normal PSD. Sometimes the effective variance

$$\Delta_{ef} = \frac{\overline{a^2(a-a_{ef})^2}}{\overline{a^2}a_{ef}^2} \qquad (1.7)$$

is also used. Note that $\Delta_{ef} = (3+\mu)^{-1}$ for the gamma distribution and $\Delta_{ef} = \exp(\sigma^2) - 1$ for the log-normal distribution. The values of a_{ef} and Δ for the log-normal and gamma particle size distributions coincide, if

$$\sigma = \sqrt{\ln\frac{2+\mu}{1+\mu}}, \qquad a_m = \left(\frac{2+\mu}{1+\mu}\right)^{2/5}\left(1+\frac{3}{\mu}\right)a_0. \qquad (1.8)$$

It follows from Equations 1.8 that as $\mu \to \infty$ (monodispersed media), $\sigma \to 0$, and $a_m \to a_0$. For example, characteristic values of (a_{ef}, Δ) are (6 µm, 0.4) for water clouds and (0.25 µm, 0.6) for stratospheric aerosols.

It should be pointed out that the size and shape of particles are often correlated. For example, larger particles in ice clouds exhibit more deviation from sphericity than small crystals do.

There are different approaches to characterize the shape of light scattering particles. One is to introduce two broad classes, namely, regular and irregular particles. Examples of regular particles are spheres, spheroids, circular cylinders, cubes, tetrahedrons, and hexagonal cylinders, to name but a few. The volume and surface area of such particles can be easily calculated (see Table 1.1). Statistical methods should be used to characterize the shape of irregular particles. Examples of media with irregular shaped particles are soot aerosols, ice clouds, and oceanic suspensions.

Nonspherical particles can be characterized by the following parameters:

- the ratio of the maximal to minimal dimensions;
- the mean deviation of the surface from a smoothed particle boundary;
- the number of edges (the sharpness of the surface deviation).

A more sophisticated approach to shape characterization is based on the introduction of a three-dimensional stochastic function. Multivariate log-normal statistics can be used for this purpose. Irregularly shaped particles are often described in terms of the fractal geometry (e.g., triadic Koch fractals).

Nonspherical particles can be fully or partially oriented by gravitational, electrical, or magnetic forces and they can have different internal structures (e.g., shell structure, aggregates, 'raisin pudding').

The electromagnetic response of particles fully depends on their size, shape, orientation, concentration and internal structure. Thus, one can imagine the richness of results and possibilities that exist in the optics of disperse media. Many of them have already been exploited for practical applications, ranging from magnetic recording media to solar battery construction. Even more possibilities occur if one considers particles composed of anisotropic or chiral substances.

Table 1.1. The expressions for volumes (V), surface areas (Σ), and effective radii ($a_{ef} = 3V/\Sigma$) of particles having different shapes.

Type of particle	V	Σ	a_{ef}
Sphere (a–radius)	$\dfrac{4\pi a^3}{3}$	$4\pi a^2$	a
Prolate spheroid (b, r, and l are semi-axes, $b = r < l$, $\varepsilon = \sqrt{1-\xi^2}$, and $\xi = r/l$)	$\dfrac{4\pi}{3}r^2 l$	$4\pi r^2 \left[\dfrac{1}{2} + \dfrac{\arcsin\sqrt{1-\xi^2}}{2\xi\sqrt{1-\xi^2}}\right]$	$\dfrac{2r}{\xi + \dfrac{\arcsin\sqrt{1-\xi^2}}{\sqrt{1-\xi^2}}}$
Oblate spheroid (b, r, and l are semi-axes, $b = r > l$, $\varepsilon' = \sqrt{\xi^2 - 1}$, and $\xi = r/l$)	$\dfrac{4\pi}{3}r^2 l$	$4\pi r^2 \left[1 + \dfrac{\ln\dfrac{1+\varepsilon'}{1-\varepsilon'}}{2\xi^2\varepsilon'}\right]$	$\dfrac{r}{\xi + \dfrac{\ln\dfrac{1+\varepsilon'}{1-\varepsilon'}}{2\xi^2\varepsilon'}}$
Circular cylinder (l is the length, r is the radius, $\xi = r/l$)	$\pi r^2 l$	$2rl(1+\pi\xi)$	$\dfrac{3\pi r}{2(1+\pi\xi)}$
Cube (l is the length)	l^3	$8l^2$	$\dfrac{3}{8}l$
Tetrahedron (l is the height, a is the side of the base)	$\dfrac{la^2}{4\sqrt{3}}$	$a^2\sqrt{3}$	$\dfrac{l}{4}$
Hexagonal cylinder (l is the length, r is the side of the base, $\xi = r/l$)	$\dfrac{\sqrt{3}r^2 l}{2}$	$6rl\left(1 + \dfrac{\xi}{2\sqrt{3}}\right)$	$\dfrac{\sqrt{3}r}{4\left(1 + \dfrac{\xi}{2\sqrt{3}}\right)}$

1.2 LIGHT BEAMS

The main topic of this book is the description of the interaction of light beams with small particles. In many respects the characterization of a light beam is simpler than that of particles. It is defined by its direction, frequency range, intensity and state of polarization. In this book only low intensity beams (linear optics) are considered, though the state of polarization can be arbitrary.

The polarization of a light beam is described by different means. The most common approach to this problem was proposed by Stokes (1852). He introduced the 4-dimensional vector-parameter $\vec{S}$ with components I, Q, U, and V. These components have the same dimension. The most important property of the Stokes vector $\vec{S}$ is that when several independent (without permanent phase relations) beams of light, propagating in the same direction, are combined, the Stokes parameters, I, Q, U, V, for the mixture is the sum of respective Stokes parameters of the separate beams.

The Stokes parameters of completely polarized light are defined by the following equations:

$$I = E_1 E_1^* + E_2 E_2^*, \tag{1.9}$$

$$Q = E_1 E_1^* - E_2 E_2^*, \tag{1.10}$$

$$U = E_1 E_2^* + E_2 E_1^*, \tag{1.11}$$

$$V = i(E_1 E_2^* - E_2 E_1^*), \tag{1.12}$$

where common multipliers are omitted for the sake of simplicity. Values of E_1 and E_2 are components of the complex electric vector $\vec{E}$ in the plane perpendicular to the direction of propagation, specified by the unity vector $\vec{e}_3 = \vec{e}_2 \times \vec{e}_1$:

$$\vec{E} = E_1 \vec{e}_1 + E_2 \vec{e}_2. \tag{1.13}$$

Thus, the general property of the electromagnetic waves, $\vec{E} \perp \vec{e}_3$, is used in the definition of the Stokes vector.

Note that the light beam is a mixture of electromagnetic waves with different amplitudes and phases. Thus, Equations 1.9–1.12 should be averaged with respect to variations of the phase and amplitude of a single wave.

There is an inconvenience related to the definitions of Equations 1.9–1.12, namely, that they define beam characteristics in a fixed coordinate system, which is determined by the direction of beam propagation. Photons can change their directions many times due to scattering processes and one should account for the rotation of the coordinate system at each point where scattering occurs.

This problem can be avoided if one uses the notion of the light beam tensor (Fedorov, 1958, 1976):

$$\hat{\Phi} = \begin{pmatrix} E_1 E_1^* & E_1 E_2^* & E_1 E_3^* \\ E_2 E_1^* & E_2 E_2^* & E_2 E_3^* \\ E_3 E_1^* & E_3 E_2^* & E_3 E_3^* \end{pmatrix}, \tag{1.14}$$

where $\vec{E} = E_1 \vec{i} + E_2 \vec{j} + E_3 \vec{k}$ is the electromagnetic field vector defined in the rectangular coordinate system, determined by vectors $\vec{i}, \vec{j},$ and $\vec{k}$. Note that the definition of the tensor $\hat{\Phi}$ is not related to the specific direction of a light beam and one does not need to make rotations of the coordinate system after each scattering process. This allows for introduction of tensor covariant methods into radiative transfer theory and optics of light scattering media (see Appendix 1).

Electromagnetic waves oscillate only in the direction perpendicular to the propagation direction. Therefore, we have:

$$E_3 = 0 \tag{1.15}$$

at $\vec{e}_3 \| \vec{k}$ ($\vec{e}_3$ determines the direction of propagation). This allows for the reduction of the light beam tensor (Equation 1.14) to the density matrix $\hat{\rho}$, defined as

$$\hat{\rho} = \begin{pmatrix} E_1 E_1^* & E_1 E_2^* \\ E_2 E_1^* & E_2 E_2^* \end{pmatrix} \tag{1.16}$$

or in short notation: $\hat{\rho} = \vec{E} \otimes \vec{E}^+$, where '$\otimes$' means the direct product and '+' means the simultaneous operation of transportation and conjugation. Clearly, the description of the light beam using $\hat{\rho}$ is equivalent to the description with the Stokes parameters (Equations 1.9–1.12). In particular, we have:

$$I = \rho_{11} + \rho_{22}, \quad Q = \rho_{11} - \rho_{22}, \quad U = \rho_{12} + \rho_{21}, \quad V = i(\rho_{12} - \rho_{21}). \quad (1.17)$$

Let us note the following relation:

$$\hat{\rho} = \tfrac{1}{2}[I\hat{\sigma}_0 + Q\hat{\sigma}_1 + U\hat{\sigma}_2 + V\hat{\sigma}_3], \quad (1.18)$$

where

$$\hat{\sigma}_0 = \begin{pmatrix} 1 & 0 \\ 0 & 1 \end{pmatrix}, \hat{\sigma}_1 = \begin{pmatrix} 1 & 0 \\ 0 & -1 \end{pmatrix}, \hat{\sigma}_2 = \begin{pmatrix} 0 & 1 \\ 1 & 0 \end{pmatrix}, \hat{\sigma}_3 = \begin{pmatrix} 0 & -i \\ i & 0 \end{pmatrix} \quad (1.19)$$

are linearly independent spin matrices. Therefore, the Stokes parameters are proportional to the coefficients of the expansion of $\hat{\rho}$ in terms of the spin matrices $\hat{\sigma}_j$.

It should be pointed out that light scattering theory and radiative transfer can be formulated both in terms of the Stokes parameters and the light beam tensor. The Stokes formulation simplifies solutions for isotropic media. However, a coordinate-free approach is more suitable for the solution of problems in the optics of light scattering anisotropic media.

It is known that the most general oscillation of the electric vector of a plane wave in the plane perpendicular to the direction of propagation is of an elliptic type. The characteristics of the polarization ellipse, namely the azimuth ψ (the position of the major axis relative to the vector $\vec{e}_1$) and the ellipticity $e \leq 1$ (ratio of axes), can be derived from the Stokes parameters of a beam using the following equations (Bohren and Huffman, 1983):

$$\psi = \tfrac{1}{2}\arctan\frac{U}{Q}, \quad e = |\tan\varphi|, \quad (1.20)$$

where

$$\varphi = \tfrac{1}{2}\arcsin\frac{V}{w} \quad (1.21)$$

is the ellipticity angle and $w = \sqrt{U^2 + Q^2 + V^2}$. The value of

$$q = \frac{w}{I} \quad (1.22)$$

is called the degree of polarization of the light beam. It follows for unpolarized radiation that $U = Q = V = 0$ and $q = 0$. Values ψ, e, φ in Equations 1.20 and 1.21 are not defined in this case. One can also introduce the degree of circular polarization $p_c = V/I$ and the degree of linear polarization $p_l = (\sqrt{U^2 + Q^2}/I)$. Then $q = \sqrt{p_l^2 + p_c^2}$.

The dependence of the azimuth ψ on the ratio U/Q (see Equation 1.20) is presented in Figure 1.1. One can see that there are two values of the angle ψ for the same ratio U/Q. One must choose that value which gives $\cos 2\psi$ the same sign as Q. This means that the central line in Figure 1.1 should be used only for negative

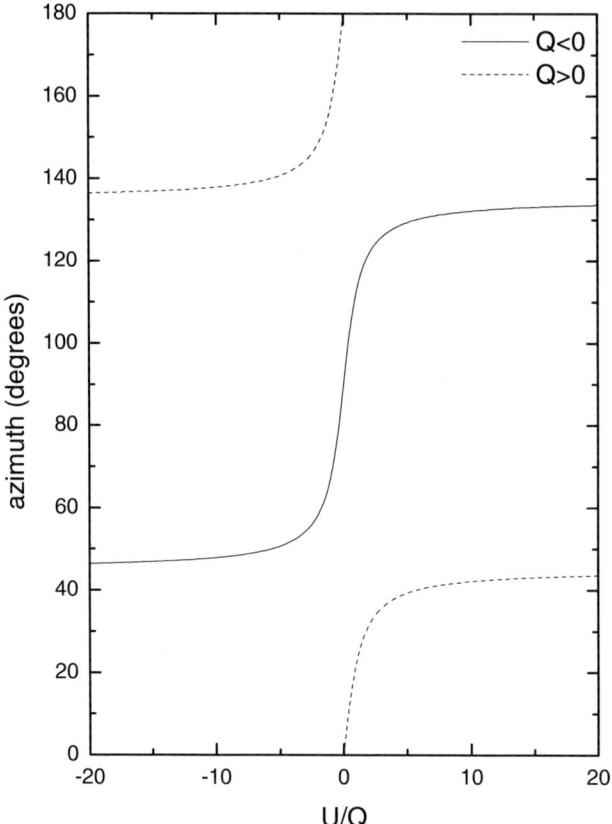

Figure 1.1. The dependence of the azimuth ψ on the ratio U/Q.

values of the parameter Q. The dependence of the ellipticity angle φ on the ratio V/w (see Equation 1.21) is presented in Figure 1.2. One can see that $\varphi \approx V/2w$ at $|V/w| \leq 0.4$. The positive values of φ mean that the polarization is right-handed or the electric vector traces the polarization ellipse in the clockwise sense when looking in the direction from which the light is coming. The dependence of the ellipticity e on the angle φ is presented in Figure 1.3. We see that $e \approx \varphi$ at $|\varphi| \leq 0.2$. Note that it follows that $\psi \in [0, \pi]$, $\varphi \in [-\pi/4, \pi/4]$, $e \in [0, 1]$.

It is important that the Stokes vector of the partially polarized light beam $\vec{S}(I, Q, U, V)$ can be represented as a sum of the Stokes vectors of the unpolarized beam:

$$\vec{S}_u = ((1-q)I, 0, 0, 0) \tag{1.23}$$

and the Stokes vector of the completely polarized beam

$$\vec{S}_p = (qI, Q, U, V). \tag{1.24}$$

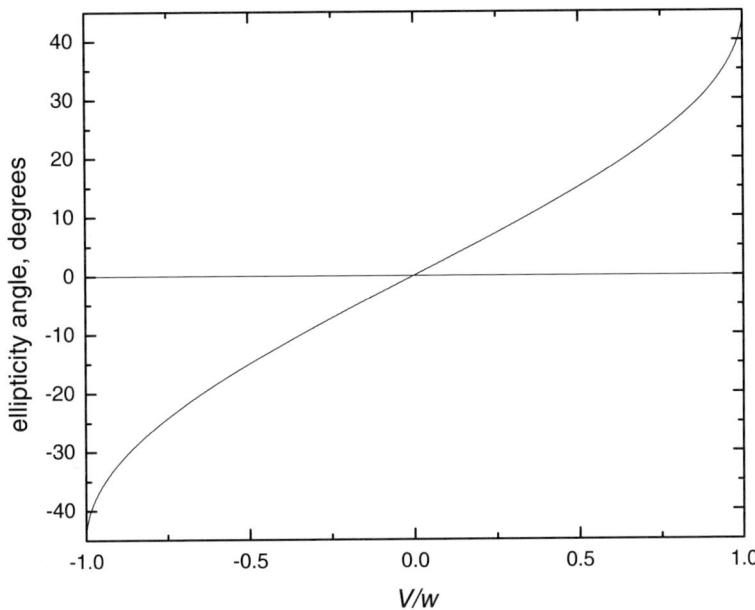

Figure 1.2. The dependence of the ellipticity angle φ on the ratio V/w.

We underline that it is possible to find the light intensity I, the degree of polarization q, and the characteristics of the polarization ellipse (φ, ψ) from the easily measurable Stokes parameters. There are no polarization measurements which can distinguish between light beams with the same values of the Stokes vector $\vec{S}$ (or sets (I, q, φ, ψ), (I, p_l, p_c, ψ)).

Note that the parameters I, V, q, p_l, p_c, and φ are coordinate independent. This is not the case for U, Q, and ψ. For instance, in atmospheric optics applications, the angle ψ is usually defined with respect to the meridional plane, which contains the normal to a plane-parallel light scattering layer and the observation direction $\vec{n}$. Oscillations in the electromagnetic wave, propagating in the direction $\vec{n}$, occur in the plane perpendicular to $\vec{n}$. In particular, the electric vector $\vec{E}$ could lie in the meridional plane ($\psi = 0$), be perpendicular to this plane ($\psi = \pi/2$), or have any other value in the range $[0, \pi]$. If we choose another plane as a reference plane, then ψ also changes. Remember that light intensity I and degrees of polarization q, p_l, p_c in the direction $\vec{n}$ are invariants of such a transformation. In single scattering problems the angle ψ is usually measured with respect to the scattering plane which contains incident and scattered light beams.

We will use the Stokes vector-parameter formulation for the characterization of light beam transformations during its propagation and scattering in a disperse medium. This approach is much simpler than that based on the field characteristics. Note that only directly measurable optical quantities are considered in the framework of the Stokes formulation. The vector $\vec{E}$ itself cannot be measured in the optical band. This is an important difference to the microwave region.

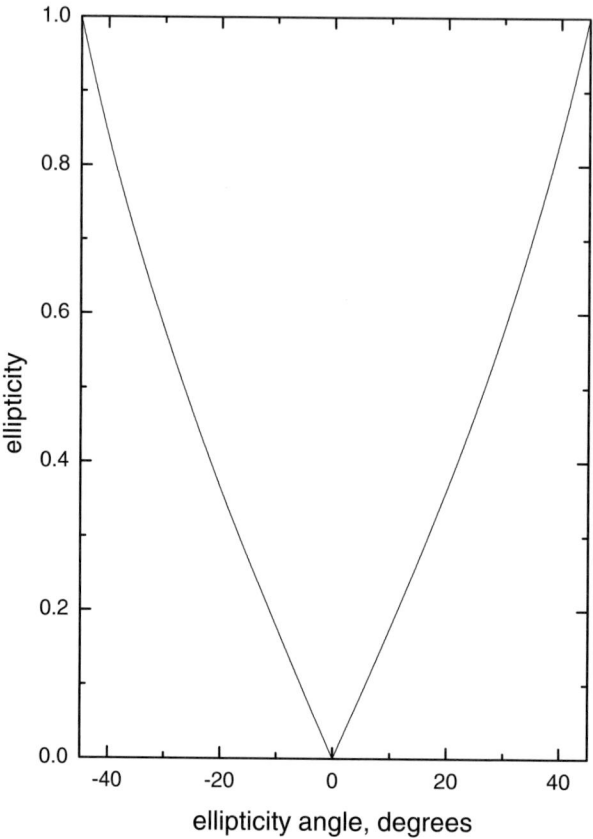

Figure 1.3. The dependence of the ellipticity e on the ellipticity angle φ.

1.3 THE INTERACTION MATRIX

The linear interaction of an electromagnetic field with a single particle or an elementary volume of a disperse medium can be described by the following general equation:

$$\vec{E} = \hat{A}\vec{E}_0. \qquad (1.25)$$

Here $\vec{E}_0$ is the electric vector of the incident wave and $\vec{E}$ is the electric vector of the wave after the interaction event. The interaction matrix $\hat{A}$ can be calculated, using Maxwell's electromagnetic theory.

We, however, prefer to deal with bilinear forms (e.g., $\hat{\rho}$ and $\vec{S}$), which can be directly measured. This is not the case for the vector $\vec{E}$ in the optical range, due to its high frequency oscillations of the electromagnetic field in the light beam (typically, 10^{15} oscillations per second). Optical detectors average measurements in both time and space domains, and theoretical descriptions should be consistent with measurement techniques.

Let us establish the transformation law for the density matrix

$$\hat{\rho} = \vec{E} \otimes \vec{E}^+ \tag{1.26}$$

due to a scattering event. The substitution of Equation 1.25 into Equation 1.26 gives:

$$\hat{\rho} = \hat{A}\vec{E}_0 \otimes (\hat{A}\vec{E}_0)^+ \tag{1.27}$$

or

$$\hat{\rho} = \hat{A}\hat{\rho}_0\hat{A}^+, \tag{1.28}$$

where

$$\hat{\rho}_0 = \vec{E}_0 \otimes \vec{E}_0^+ \tag{1.29}$$

is the density matrix of the incident light beam. Equation 1.28 gives the transformation law required. It can be used for the determination of the interaction matrix $\hat{M}$ in the Stokes vector formulation:

$$\vec{S} = \hat{M}\vec{S}_0, \tag{1.30}$$

where $\vec{S}_0$ is the Stokes vector of the incident light and $\vec{S}$ is the Stokes vector of the light beam after the interaction event. Let us find relationships between matrices $\hat{A}$ and $\hat{M}$.

It follows from Equation 1.18 that

$$S_j = Tr(\hat{\sigma}_j\hat{\rho}), \tag{1.31}$$

where we used the following property of spin matrices: $Tr(\hat{\sigma}_q\hat{\sigma}_p) = 2\delta_{qp}$. Here Tr means the trace operation and δ_{qp} is the Kronecker symbol. In the quantum mechanical language, Equation 1.31 means that S_j gives the expectation value of the matrix $\hat{\sigma}_j$ in the polarization state represented by the density matrix $\hat{\rho}$. Substituting Equation 1.28 in Equation 1.31, we obtain:

$$S_j = Tr(\hat{\sigma}_j\hat{A}\hat{\rho}_0\hat{A}^+). \tag{1.32}$$

It follows from Equation (1.18):

$$\hat{\rho}_0 = \frac{1}{2}\sum_{k=0}^{3} S_{0k}\hat{\sigma}_k, \tag{1.33}$$

where S_{0k} are components of the Stokes vector of the incident light field. The substitution of Equation 1.33 in Equation 1.32 gives:

$$S_j = \frac{1}{2}Tr\left(\sum_{k=0}^{3}\hat{\sigma}_j\hat{A}\hat{\sigma}_k\hat{A}^+\right)S_{0k}, \tag{1.34}$$

where the scalar multiplier S_{0k} is taken out of the trace operation. Finally, comparing Equations 1.30 and 1.34, we obtain for the elements of the matrix $\hat{M}$:

$$M_{jk} = \frac{1}{2}Tr(\hat{\sigma}_j\hat{A}\hat{\sigma}_k\hat{A}^+). \tag{1.35}$$

Here the indices j and k take values 0, 1, 2, 3. This matrix is of paramount importance for studies of arbitrarily polarized light beam interactions with disperse media. It is called the Mueller matrix and will be considered in detail later on for various types of scattering particles.

The Mueller matrix is completely determined by the interaction matrix $\hat{A}$. There are only 7 independent numbers, which characterize the complex $2*2$ matrix $\hat{A}$ (the relative phase is of no importance here). This means that there are nine relationships between 16 elements of the $4*4$ Mueller matrix $\hat{M}$. Matrices $\hat{M}$ can be introduced for any media, including multiple light scattering ones. Some examples are given in Appendix 2. We also give examples of the interaction matrix $\hat{A}$ (which is often called the Jones matrix) and vectors $\vec{S}$ for various polarized light beams there.

1.4 FURTHER READING

Allen, T. (1990) *Particle Size Measurement*. Chapman & Hall, London.
Bohren, C. F. and Huffman, D. R. (1983) *Absorption and Scattering of Light by Small Particles*. Wiley, New York.
Brosseau, C. (1998) *Fundamentals of Polarization Optics: A Statistical Approach*. Wiley, New York.
Chandrasekhar, S. (1950) *Radiative Transfer*. Oxford University Press, Oxford, UK.
Collett, E. (1993) *Polarized Light: Fundamentals and Applications*. Marcel Dekker, New York.
Fedorov, F. I. (1958) *Optics of Anizotropic Media*. Nauka and Tekhnika, Minsk.
Fedorov, F. I. (1976) *Theory of Gyrotropy*. Nauka and Tekhnika, Minsk.
Jauch, J. M. and Rohrlich, F. (1955) *Theory of Photons and Electrons*. Addison-Wesley, Reading, MA.
Mishchenko, M. I., Travis, L. D., and Lacis, A. A. (2002) *Scattering, Absorption, and Emission of Light by Small Particles*. Cambridge University Press, Cambridge, UK.
Perrin, F. (1942) Polarization of light scattered by isotropic opalescent media. *J. Chem. Phys.*, **10**, 415–427.
Rozenberg, G. V. (1955) Stokes vector-parameter. *Uspekhi Fiz. Nauk*, **56**, 77–110.
Silverman, M. (1998) *Waves and Grains*. Princeton University Press, Princeton, NJ.
Stokes, G. G. (1852) On the composition and resolution of streams of polarized light from different sources. *Trans. Camb. Phil. Soc.*, **9**, 339–416.
Van de Hulst, H. C. (1957) *Light Scattering by Small Particles*. Wiley, New York.
Van de Hulst, H. C. (1980) *Multiple Light Scattering: Tables, Formulas and Applications*. Academic Press, New York.

2

Polarized radiative transfer

2.1 VECTOR RADIATIVE TRANSFER EQUATION

The interaction of a light beam with a disperse medium of an arbitrary thickness is usually described in the framework of the radiative transfer theory, which is based on the solutions of the integro-differential radiative transfer equation (Chandrasekhar, 1950). This equation can be easily derived, using the energy conservation law.

We consider a light beam propagated in an absorbing, emitting, and scattering disperse medium. The intensity I of this beam will be constant in the absence of emission, absorption, scattering, and refraction of light in the medium. It changes with distance in disperse media, however.

The change of the intensity dI after traversing the thickness dL in the direction specified by a unit vector $\vec{\Omega}$ in the linear optics approximation is given by the following formula:

$$dI(\vec{\Omega}) = -\sigma_{ext} I(\vec{\Omega}) dL + B_{em}(\vec{\Omega}) dL \qquad (2.1)$$

where σ_{ext} is the extinction coefficient and the function $B_{em}(\vec{\Omega})$ describes the emission and scattering from all directions to the direction given by $\vec{\Omega}$. We neglect possible frequency change in the scattering process. It is evident that the decrease of the intensity is caused by the removal of photons, propagating in the fixed direction $\vec{\Omega}(\vartheta, \phi)$, due to their absorption and scattering in other directions (ϑ and ϕ are the zenith and azimuth angles, respectively). This process is described by the first term in Equation 2.1. Thus, the value of σ_{ext} can be divided into two parts:

$$\sigma_{ext} = \sigma_{abs} + \sigma_{sca}, \qquad (2.2)$$

where σ_{abs} and σ_{sca} are absorption and scattering coefficients. These coefficients depend on the number concentration of particles N and their absorption C_{abs} and scattering C_{sca} cross sections. In particular, we have for monodisperse spheres:

$$\sigma_{ext} = NC_{sca}, \sigma_{abs} = NC_{abs}. \qquad (2.3)$$

The extinction cross section is defined as a sum of the scattering and absorption cross sections, namely, $C_{ext} = C_{sca} + C_{abs}$. For particles of different sizes, shapes, or chemical compositions one should use mean values $\langle C_{sca} \rangle$ and $\langle C_{abs} \rangle$ in Equation 2.3. In the case of uniform spheres, it follows that

$$\langle C_{sca} \rangle = \int_0^\infty f(a) C_{sca}\, da, \qquad (2.4)$$

$$\langle C_{abs} \rangle = \int_0^\infty f(a) C_{abs}\, da, \qquad (2.5)$$

where $f(a)$ is the particle size distribution.

In fact, simple linear relationships between the parameters of single particles (see for example Equation 2.3) and those of the radiative transfer equation (RTE) (Equation 2.1) are valid only for disperse media with low volumetric concentrations of scatterers $c_v = NV < 0.01$–0.1. Here V is the average volume of particles in a unit volume of a scattering medium. One should account for the correlation of particles positions in the case of high concentrations.

Moreover, the whole approach, based on the RTE (2.1) may become invalid for close-packed media (Apresyan and Kravtsov, 1983). For instance, spherical particles at volumetric concentrations $c_v \approx 0.74$ form almost perfectly ordered structures. Scattering of light by such structures is closer to the diffraction of light by crystalline media. For instance, Bragg maxima appear. Clearly, highly ordered particulate media can not be described in the framework of the standard radiative transfer theory.

Fortunately, almost all natural media (cosmic dust, atmospheric aerosols, clouds, oceanic suspensions) are characterized by extremely low values of the volumetric concentration c_v. Thus, the linear approximation (Equation 2.3) can be applied in this case. For instance, characteristic values of c_v are in the range 10^{-11}–10^{-7} for water clouds.

Let us return to Equation 2.1 now. The value of B_{em} in this equation consists of two parts:

$$B_{em} = B'_{em} + B''_{em} \qquad (2.6)$$

where the value of B'_{em} is due to the internal sources of radiation inside a medium and the value of B''_{em} accounts for photons scattered from other directions to the direction specified by $\vec{\Omega} = (\vartheta, \phi)$. It is evident that (Chandrasekhar, 1950)

$$B'_{em} = B(T) \qquad (2.7)$$

for media in the local thermodynamic equilibrium. Here,

$$B(T) = \frac{2h\nu^3}{c^2} \frac{1}{e^{h\nu/kT} - 1}$$

is the Planck's function, ν is the frequency, c is the speed of light, T is the temperature, h and k are Boltzmann and Planck constants, respectively.

For the value of B''_{em} it follows that

$$B''_{em} = \int_{4\pi} \sigma^d_{sca}(\vec{\Omega}, \vec{\Omega}') I(\vec{\Omega}') \, d\vec{\Omega}' \tag{2.8}$$

where $\sigma^d_{sca}(\Omega, \Omega')$ is the differential scattering cross section from the direction $\vec{\Omega}' = (\vartheta', \phi')$ to the direction $\vec{\Omega} = (\vartheta, \phi)$. The integral term (Equation 2.8) describes the scattering from all possible directions $\vec{\Omega}' = (\vartheta', \phi')$ to the observation direction $\vec{\Omega} = (\vartheta, \phi)$. Summing up, the RTE (2.1) can be represented as follows:

$$\frac{dI(\vartheta, \phi)}{dL} = -\sigma_{ext} I(\vartheta, \phi) + \int_0^{2\pi} d\phi' \int_0^{\pi} d\vartheta' \sin\vartheta' \sigma^d_{sca}(\vartheta, \vartheta', \phi, \phi') I(\vartheta', \phi'), \tag{2.9}$$

where we have neglected the term B''_{em}, which is often of no importance in the visible region of the electromagnetic spectrum ($B(T) \to 0$ at $\nu \to \infty$). For isotropic media the value of $\sigma^d_{sca}(\vartheta, \vartheta', \phi, \phi')$ depends on the scattering angle θ and not separately on $\vartheta, \vartheta', \varphi, \varphi'$. In particular, we have:

$$\cos\theta = uu' + \sqrt{(1-u^2)(1-u'^2)} \cos(\phi - \phi'), \tag{2.10}$$

where $u = \cos\vartheta$, $u' = \cos\vartheta'$.

One can see that the microphysical parameters of a disperse medium (e.g., $f(a)$) enter the RTE (2.9) via the extinction coefficient σ_{ext} and the differential scattering coefficient σ^d_{sca} or alternatively via values of the optical depth $\tau = \sigma_{ext} L$ and the ratio $\Psi = \sigma^d_{sca}/\sigma_{ext}$. Thus, media with the same values of τ and Ψ are characterized by exactly the same radiative characteristics. This feature allows one to model radiative transfer in terrestrial clouds of different shapes in the laboratory, avoiding expensive field measurements. For this, one should choose the value τ in a laboratory experiment close to the value of this product for a cloudy medium (e.g., $\tau = 30$). It could be done, for example, by reducing the value of L, which could be 1 km or longer for natural clouds, and increasing σ_{ext} correspondingly. The function Ψ, the source of light and the receiver also should be modelled to simulate an experimental set-up in the real atmosphere. Of course, the same approach can be applied to other types of media (e.g., oceanic water).

The scattering coefficient σ_{sca} is defined as an integral of the differential cross section σ^d_{sca}, namely:

$$\sigma_{sca} = \int_0^{2\pi} d\varphi \int_0^{\pi} \sigma^d_{sca}(\theta, \varphi) \sin\theta \, d\theta \tag{2.11}$$

or

$$\sigma_{sca} = 2\pi \int_0^{\pi} \sigma^d_{sca}(\theta) \sin\theta \, d\theta \tag{2.12}$$

for media with an azimuthally independent local light scattering law. It follows from Equation 2.12 that

$$\int_0^{\pi} \frac{2\pi \sigma^d_{sca}(\theta)}{\sigma_{sca}} \sin\theta \, d\theta = 1. \tag{2.13}$$

The value

$$p(\theta) = \frac{4\pi \sigma_{sca}^d(\theta)}{\sigma_{sca}} \qquad (2.14)$$

is called the phase function or the scattering indicatrix. The phase function is normalized by the following condition (see Equations 2.13 and 2.14):

$$\frac{1}{2}\int_0^\pi p(\theta) \sin\theta \, d\theta = 1. \qquad (2.15)$$

Note that the probability of photon scattering from the direction $\vec{\Omega}' = (\vartheta', \phi')$ to the direction $\vec{\Omega} = (\vartheta, \phi)$ is equal to $p(\theta)\,d\Omega/4\pi$. Thus, the notion of the phase function can be used for a probabilistic interpretation of the RTE (Sobolev, 1956).

The differential scattering cross section does not depend on the scattering angle in so-called isotropically scattering media. Thus, it follows in this case that $\sigma_{sca}^d = K$, where $K = $ const. The total scattering cross section can be obtained from Equation 2.12: $\sigma_{sca} = 4\pi K$. The phase function is equal to 1 in this case and the probability of photon scattering within the solid angle $d\Omega$ is equal to $d\Omega/4\pi$.

Only radiative transfer in plane-parallel random media is considered in this book. Then linear distances can be measured along the normal to the plane of the stratification. It follows from Equation 2.9 for a particular case of a plane-parallel layer:

$$\cos\vartheta\,\frac{dI(\vartheta,\phi)}{dZ} = -\sigma_{ext}I(\vartheta,\phi) + \frac{\sigma_{sca}}{4\pi}\int_0^{2\pi}d\phi'\int_0^\pi d\vartheta' \sin\vartheta'\,p(\cos\theta)I(\vartheta',\phi'). \qquad (2.16)$$

Here, the axis $\vec{OZ}$ (directed downward) is perpendicular to the boundary of the medium (see Figure 2.1), $\cos\theta = \cos\vartheta\cos\vartheta' + \sin\vartheta\sin\vartheta'\cos(\phi-\phi')$ is the cosine of the scattering angle and $Z = L\cos\vartheta$. The values of ϑ and ϑ' denote zenith angles with the downward vertical, ϕ and ϕ' are azimuths referred to a chosen x-axis and

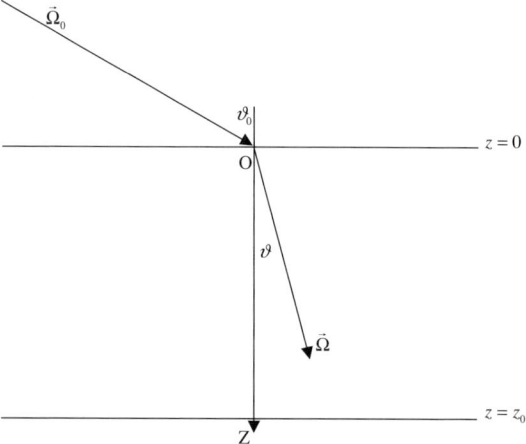

Figure 2.1. The geometry of the problem. A scattering layer is placed between planes $z = 0$ and $z = z_0$. The vector $\vec{\Omega}$ gives the direction of illumination.

measured clockwise when looking upward. This equation can be applied to radiative transfer problems in isotropic plane parallel media with discrete particles.

The question arises about the possibility of extending the formulation (Equation 2.16) to account for the transformation of the light beam polarization characteristics during its propagation and multiple light scattering in a disperse medium. This is done quite easily, using the Stokes vector formulation introduced above. Then we have (Chandrasekhar, 1950; Rozenberg, 1955):

$$\frac{d\vec{S}(\vartheta,\phi)}{dL} = -\hat{\sigma}_{ext}\vec{S}(\vartheta,\phi) + \int_0^{2\pi} d\phi' \int_0^{\pi} d\vartheta' \sin\vartheta' \hat{\sigma}_{sca}^d(\vartheta,\vartheta',\phi,\phi')\vec{S}(\vartheta',\phi'), \quad (2.17)$$

where $\vec{S} = (I, Q, U, V)$ is the Stokes vector, $\hat{\sigma}_{ext}$ is the extinction matrix and $\hat{\sigma}_{sca}^d$ is the differential scattering matrix. Note, that it follows for plane parallel media (axis $\vec{OZ}$ is perpendicular to the boundary of a medium):

$$\cos\vartheta \frac{d\vec{S}(\theta,\phi)}{dZ} = -\hat{\sigma}_{ext}\vec{S}(\theta,\phi)$$
$$+ \int_0^{2\pi} d\phi' \int_0^{\pi} d\vartheta' \sin\vartheta' \hat{\sigma}_{sca}^d(\vartheta,\vartheta',\phi,\phi')\vec{S}(\theta',\phi'), \quad (2.18)$$

where the Stokes vector $\vec{S}(\theta,\phi)$ is defined with respect to the meridional plane, which holds directions $\vec{OZ}$ and $\vec{\Omega} = (\vartheta,\phi)$. Note that scattering matrices $\hat{\sigma}_{sca}(\theta)$ in the single light scattering theory are usually calculated in respect to the scattering plane, which holds directions $\vec{\Omega}' = (\vartheta',\phi')$ and $\vec{\Omega} = (\vartheta,\phi)$ (Van de Hulst, 1981). Thus, one needs to make rotations of the matrix $\hat{\sigma}_{sca}$, defined in the single scattering theory, to obtain the matrix $\hat{\sigma}_{sca}^d$ in Equation 2.18. Namely, it follows:

$$\hat{\sigma}_{sca}^d = \hat{H}(\pi - i_2)\hat{\sigma}_{sca}(\theta)\hat{H}(-i_1), \quad (2.19)$$

where according to simple trigonometric calculations (Hovenier, 1971):

$$\hat{H}(-i_1) = \begin{pmatrix} 1 & 0 & 0 & 0 \\ 0 & \cos 2i_1 & -\sin 2i_1 & 0 \\ 0 & \sin 2i_1 & \cos 2i_1 & 0 \\ 0 & 0 & 0 & 1 \end{pmatrix},$$

$$\hat{H}(\pi - i_2) = \begin{pmatrix} 1 & 0 & 0 & 0 \\ 0 & \cos 2i_2 & -\sin 2i_2 & 0 \\ 0 & \sin 2i_2 & \cos 2i_2 & 0 \\ 0 & 0 & 0 & 1 \end{pmatrix}, \quad (2.20)$$

$$\cos 2i_j = 2\cos^2 i_j - 1, \sin 2i_j = 2\sqrt{1-\cos^2 i_j}\cos i_j,$$

$$\cos i_1 = \frac{-u + u'\cos\theta}{s\sqrt{(1-\cos^2\theta)(1-u'^2)}}, \quad \cos i_2 = \frac{-u' + u\cos\theta}{s\sqrt{(1-\cos^2\theta)(1-u^2)}}$$

and $s = \text{sgn}(\phi - \phi' - \pi), j = 1, 2.$

The extinction matrix $\hat{\sigma}_{ext}$ in Equation 2.18 is reduced to the scalar value σ_{ext} for isotropic symmetric media. It follows from Equation 2.18 in this case:

$$\cos\vartheta \frac{d\vec{S}(\vartheta,\phi)}{d\tau} = -\vec{S}(\vartheta,\phi) + \frac{\omega_0}{4\pi}\int_0^{2\pi}d\phi'\int_0^\pi d\vartheta'\sin\vartheta'\hat{F}(\vartheta,\phi,\vartheta',\phi')\vec{S}(\vartheta',\phi'), \quad (2.21)$$

where

$$\left.\begin{array}{l}\hat{F}(\vartheta,\phi,\vartheta',\phi') = \hat{H}(\pi-i_2)\hat{P}(\vartheta,\phi,\vartheta',\phi')\hat{H}(-i_1),\\[6pt]\hat{P}(\vartheta,\phi,\vartheta',\phi') = \dfrac{4\pi\hat{\sigma}_{sca}(\vartheta,\vartheta',\phi,\phi')}{\sigma_{sca}}.\end{array}\right\} \quad (2.22)$$

Here $\tau = \sigma_{ext}Z$ is the optical depth, Z is the geometrical depth of a light-scattering layer, $\omega_0 = \sigma_{sca}/\sigma_{ext}$ is the single scattering albedo. Matrices $\hat{F}$ and $\hat{P}$ are called the phase matrices of a light scattering medium. The matrix $\hat{P}$ is defined with respect to the scattering plane and the matrix $\hat{F}$ is given with respect to the meridional plane. Sometimes the matrix $\hat{P}$ is referred to as a scattering matrix and the matrix $\hat{F}$ as the Stokes matrix. They are related to each other by Equation (2.22). We will also use the normalized phase matrix $\hat{p} = \hat{P}/P_{11}$, where P_{11} is the first element of the matrix $\hat{P}$.

Note that the phase matrix $\hat{P}(\vartheta,\phi,\vartheta',\phi')$ of randomly oriented particles with a plane of symmetry does not change if one performs one of the following operations: interchanging ϕ and ϕ'; interchanging u and u'; or changing the sign of u and u' at once. These operations can be performed successively in an arbitrary order. Thus, the number of operations is equal to seven in total. This allows us to obtain seven symmetry relationships for the phase matrix (Hovenier, 1969). Their number in the case of randomly oriented particles having no plane of symmetry is reduced to just three.

Symmetry relationships can be used to find the symmetries in light that emerges at the top and bottom of a plane-parallel light scattering slab. In this way Hovenier (1969) proved that light incident at the top of a homogeneous layer with randomly oriented particles having the symmetry plane is equivalent to light incident at the bottom only when the azimuth is counted in the reversed sense. He pointed out that the mathematical expressions of the second and third principles of invariance for polarized light, proposed by Chandrasekhar (1950), should be changed to account for this feature.

2.2 DIRECT LIGHT

The Stokes vector $\vec{S}(\vec{\Omega})$ in Equation (2.18) can be presented as a sum of two parts:

$$\vec{S}(\vec{\Omega}) = \vec{I}(\vec{\Omega}) + \vec{I}_c(\vec{\Omega})\delta(\vec{\Omega} - \vec{\Omega_0}), \quad (2.23)$$

where the Stokes vector $\vec{I}(\vec{\Omega})$ describes an incoherent or diffused light beam and $\vec{I}_c(\vec{\Omega})\delta(\vec{\Omega} - \vec{\Omega_0})$ is the Stokes vector of the coherent or direct beam propagated in the direction $\vec{\Omega}_0(\vartheta_0, \varphi_0)$ (see Figure 2.1). Here $\delta(\vec{\Omega} - \vec{\Omega_0})$ is the delta function.

Let us substitute Equation 2.23 into Equation 2.18. Then we have two separate equations, namely, one equation for the diffused light beam – Stokes vector $\vec{I}(\vartheta, \varphi)$ and another one is for the direct (or coherent) light beam – Stokes vector $\vec{I}_c(\vartheta_0, \varphi)$:

$$\cos\vartheta \frac{d\vec{I}(\vartheta, \phi)}{dZ} = -\hat{\sigma}_{ext}\vec{I}(\vartheta, \phi) + \int_0^{2\pi} d\phi' \int_0^{\pi} d\vartheta' \sin\vartheta' \hat{\sigma}_{sca}^d(\vartheta, \vartheta', \phi, \phi')\vec{I}(\vartheta', \phi')$$
$$+ \hat{\sigma}_{sca}^d(\vartheta_0, \phi_0, \vartheta, \phi)\vec{I}_c(\vartheta_0, \phi_0), \tag{2.24}$$

$$\cos\vartheta_0 \frac{d\vec{I}_c(\vartheta_0, \phi_0)}{dZ} = -\hat{\sigma}_{ext}\vec{I}_c(\vartheta_0, \phi_0). \tag{2.25}$$

Such a separation of terms is a standard procedure in the radiative transfer theory. This allows for the simplification of both numerical and approximate analytical solutions of the RTE.

In particular, Equation 2.25 can be easily solved analytically. Then the solution $\vec{I}_c(\vartheta_0, \phi_0)$ is substituted into Equation 2.24, which can be solved using various numerical techniques. This gives the Stokes vector of the diffused light both outside and inside light scattering medium.

Let us consider the direct beam, described by Equation 2.25 in more detail. It follows from Equation 2.25:

$$\vec{I}_c = \hat{M}\vec{J}, \tag{2.26}$$

where

$$\hat{M} = \exp(-\hat{\sigma}_{ext}l), \tag{2.27}$$

where $l = Z/\cos\vartheta_0$, Z is the coordinate along the axis OZ (see Figure 2.1), and $\vec{J}$ is the Stokes vector of the incident light. We see that the vector $\vec{I}_c$ can be easily found, if the matrix $\hat{\sigma}_{ext}$ is known. In particular, the elements of the Mueller matrix $\hat{M}$ can be calculated, using Equation 1.35 and the solution for the interaction matrix, given by Ishimaru and Yeh (1984). This matrix relates the electric vectors of incident $\vec{E}_0$ and coherent fields $\vec{E}_c$, namely

$$\vec{E}_c = \hat{C}\vec{E}_0,$$

where $\hat{C}$ is the interaction matrix, which is analogous to matrix $\hat{A}$ in Equation 1.35.

We can obtain for elements of the Mueller matrix $\hat{M}$ (see Equation 1.35) (Ishimaru and Yeh, 1984):

$$M_{jk} = \tfrac{1}{2}Tr(\hat{\sigma}_j\hat{C}\hat{\sigma}_k\hat{C}^+), \tag{2.28}$$

$$\hat{C} = \hat{D}_1 e^{-\gamma_1 l} + \hat{D}_2 e^{-\gamma_2 l}, \tag{2.29}$$

where

$$\hat{D}_1 = \frac{1}{1 - \varsigma_{12}\varsigma_{21}} \begin{pmatrix} 1 & -\varsigma_{12} \\ \varsigma_{21} & -\varsigma_{12}\varsigma_{21} \end{pmatrix}, \quad \hat{D}_2 = \frac{1}{1 - \varsigma_{12}\varsigma_{21}} \begin{pmatrix} -\varsigma_{12}\varsigma_{21} & \varsigma_{12} \\ -\varsigma_{21} & 1 \end{pmatrix}, \tag{2.30}$$

and

$$\varsigma_{12} = \frac{B_{12}}{\gamma_2 - B_{11}}, \quad \varsigma_{21} = \frac{B_{21}}{\gamma_1 - B_{22}}, \quad (2.31)$$

$$\gamma_1 = \tfrac{1}{2}\left(B_{11} + B_{22} + \sqrt{4B_{12}B_{21} + (B_{11} - B_{22})^2}\right), \quad (2.32)$$

$$\gamma_2 = \tfrac{1}{2}\left(B_{11} + B_{22} - \sqrt{4B_{12}B_{21} + (B_{11} - B_{22})^2}\right). \quad (2.33)$$

Here $\hat{B} = 2\pi N k^{-2} \hat{S}(0)$, where N is the number of particles in a unit volume, $k = 2\pi/\lambda$ is the wavenumber, λ is the wavelength, and $\hat{S}(0)$ the amplitude scattering matrix (see Chapter 3) in the forward scattering direction.

Clearly, $\hat{S}(0)$ is reduced to a scalar number for isotropic symmetric media. Then we have:

$$\gamma = \gamma_1 = \gamma_2 = B_{11} = B_{22}, \quad \varsigma_{12} = \varsigma_{21} = B_{12} = B_{21} = 0,$$
$$\hat{c}_2 = \hat{h}\exp(-\gamma l), \quad \hat{M} = \hat{h}_4 \exp(-2\gamma l).$$

Here $\hat{h}_n$ is the unity matrix of dimension n.

For optically active spherical particles (Bohren and Huffman, 1983), it follows that

$$B_{11} = B_{22}, \quad B_{12} = -B_{21}, \quad \gamma_1 = \tfrac{1}{2}(B_{11} + iB_{12}), \quad \gamma_2 = \tfrac{1}{2}(B_{11} - iB_{12}),$$
$$\varsigma_{12} = \varsigma_{21} = i, \quad \hat{D}_1 = \hat{D}_1^+ = \frac{\hat{\sigma}_1 + \hat{\sigma}_4}{2}, \quad \hat{D}_2 = \hat{D}_2^+ = \frac{\hat{\sigma}_1 - \hat{\sigma}_4}{2}, \quad (2.34)$$

and, therefore,

$$\hat{M} = \begin{pmatrix} \cosh bl & 0 & 0 & -\sinh bl \\ 0 & \cos al & \sin al & 0 \\ 0 & -\sin al & \cos al & 0 \\ -\sinh bl & 0 & 0 & \cosh bl \end{pmatrix} \exp(-\varepsilon l), \quad (2.35)$$

where $\varepsilon = 4\pi N k^{-2} \operatorname{Re}(S_{11}(0))$, $a = 4\pi N k^{-2} \operatorname{Re}(S_{12}(0))$, $b = 4\pi N k^{-2} \operatorname{Im}(S_{12}(0))$. We used Equations 2.28, 2.29, 2.34 and the following properties of spin matrices

$$\hat{\sigma}_i^+ = \hat{\sigma}_i, \quad \hat{\sigma}_i^2 = \hat{\sigma}_0, \quad \hat{\sigma}_i \hat{\sigma}_j = i\hat{\sigma}_k, \quad Tr(\hat{\sigma}_i \hat{\sigma}_j) = 2\delta_{ij}, \quad \hat{\sigma}_i \hat{\sigma}_j = -\hat{\sigma}_j \hat{\sigma}_i \quad (2.36)$$

to derive the Mueller matrix $\hat{M}$ (Equation 2.35).

Note that Equations 2.26 and 2.27 (or, alternatively, Equations 2.26 and 2.28–2.33) allow one to study the transformation of the polarization characteristics of the direct light beam during its propagation through a disperse medium. It follows that the intensity of the direct beam exponentially decreases, and the polarization state is not altered in the case of isotropic symmetric media (Rozenberg, 1955).

2.3 DIFFUSED LIGHT

2.3.1 Thin layers

Let us now consider the polarization characteristics of the diffused light. Clearly, the polarization characteristics of the diffused light beam differ considerably from that of an incident beam due to effects of a multiple scattering of photons in the medium under study.

The Stokes vector of the diffused beam is governed by Equation 2.24, which takes the following form with account for Equation 2.26:

$$\cos\vartheta \frac{d\vec{I}(\vartheta,\phi)}{dZ} = -\hat{\sigma}_{ext}\vec{I}(\vartheta,\phi) + \int_0^{2\pi} d\phi' \int_0^{\pi} d\vartheta' \sin\vartheta' \hat{\sigma}^d_{sca}(\vartheta,\vartheta',\phi,\phi')\vec{I}(\vartheta',\phi')$$

$$+ \hat{\sigma}^d_{sca}(\vartheta_0,\phi_0,\vartheta,\phi)\hat{M}\vec{J}(\vartheta_0,\phi_0). \tag{2.37}$$

As far as boundary conditions are concerned we assume that there is no diffused light entering a plane-parallel light scattering layer from outside.

Equation 2.37 is a system of four coupled integro-differential equations. They can be used for studies of polarized diffused light transfer both in symmetric and asymmetric random media. For symmetric media, the centre of any large spherical volume is a centre of symmetry and any plane through this centre is a plane of symmetry. It is not the case for asymmetrical media, which usually have some rotatory power. An asymmetrical medium can be macroscopically isotropic in principle. The isotropy in this case is due to the isotropic distribution of small anisotropic elements.

Clearly, Equation 2.37 only can be solved using numerical techniques. A good starting point for the numerical procedure could be the solution of Equation 2.37 in the single scattering approximation, when the integral term in Equation 2.37 is neglected.

Let us consider now the diffused or incoherent part of the light field in the framework of the single scattering approximation. Then the integral term in Equation 2.37, which accounts for multiple light scattering, can be dropped, and we arrive at a system of four inhomogeneous linear differential equations:

$$\frac{d\vec{I}(x)}{dx} = -\hat{\sigma}_{ext}\vec{I}(x) + \vec{W}(x), \tag{2.38}$$

where $\vec{W}(x) = \hat{\sigma}^d_{sca}\hat{M}(x)\vec{J}$ using the substitution: $Z = x\cos\vartheta$. The elements of the matrix $\hat{M}(x)$ are given by Equation 2.27 (or Equation 2.28) with $l = vx$, where $v = \cos\vartheta/\cos\vartheta_0$. The system (Equation 2.38) can be solved analytically using standard techniques (see Appendix 3):

$$\vec{I}(x) = \hat{G}\vec{\Phi}(x), \tag{2.39}$$

where the matrix $\hat{G}$ is defined by the following equation:

$$\hat{\Upsilon} = \hat{G}^{-1}\hat{\sigma}_{ext}\hat{G}. \tag{2.40}$$

Here

$$\hat{\Upsilon} = \begin{pmatrix} \Lambda_1 & 0 & 0 & 0 \\ 0 & \Lambda_2 & 0 & 0 \\ 0 & 0 & \Lambda_3 & 0 \\ 0 & 0 & 0 & \Lambda_4 \end{pmatrix} \tag{2.41}$$

is the diagonalized extinction matrix. Correspondingly, $\Lambda_1, \Lambda_2, \Lambda_3$, and Λ_4 are eigenvalues of the extinction matrix $\hat{\sigma}_{ext}$. By definition, the columns of the 4 ∗ 4 matrix matrix $\hat{G}$ are given by the eigenvectors of the extinction matrix $\hat{\sigma}_{ext}$.

The elements of the vector $\vec{\Phi}(x)$ in Equation 2.39 are given by (see Appendix 3):

$$\Phi_i(x) = \hat{N}_{ij} \sigma^d_{sca,jp} F^i_{pk}(x) J_k, \qquad (2.42)$$

where $\hat{N} \equiv \hat{G}^{-1}$ and the summation on indices j, p, and k (from 0 to 3) is assumed. It follows for the elements of the matrix $\hat{F}^i$ (see Appendix 3):

$$F^i_{pk}(x) = \frac{1}{2} \sum_{r=0}^{3} \Psi^r_{pk} V^i_r(x), \qquad (2.43)$$

where

$$\Psi^1_{pk} = Tr[\hat{\sigma}_p \hat{D}_1 \hat{\sigma}_k \hat{D}_1^+], \qquad \Psi^2_{pk} = Tr[\hat{\sigma}_p \hat{D}_1 \hat{\sigma}_k \hat{D}_2^+], \qquad (2.44)$$

$$\Psi^3_{pk} = Tr[\hat{\sigma}_p \hat{D}_2 \hat{\sigma}_k \hat{D}_1^+], \qquad \Psi^4_{pk} = Tr[\hat{\sigma}_p \hat{D}_2 \hat{\sigma}_k \hat{D}_2^+] \qquad (2.45)$$

and

$$V^i_r(x) = \frac{\exp[-\Lambda_r v x] - \exp[-\Lambda_i x + (\Lambda_i - \Lambda_r v)\alpha]}{\Lambda_i - \Lambda_r v}. \qquad (2.46)$$

Here we introduced the parameter:

$$\alpha = \begin{cases} 0, & \vartheta < \frac{\pi}{2} \\ z_0 \sec \vartheta, & \vartheta \geq \frac{\pi}{2} \end{cases}, \qquad (2.47)$$

where z_0 is the geometrical thickness of a light scattering layer (see Figure 2.1).

Note that the ratio v is negative for the radiation propagated upwards ($\vartheta \geq \pi/2$) and it is positive otherwise ($\vartheta < \pi/2$).

It follows from Equation (2.46) for the reflected light ($\vartheta > \pi/2$) at the upper boundary ($x = 0, \alpha = \ell = z_0 \sec \vartheta$) of a layer:

$$V^i_r(0) = \frac{1 - \exp[(\Lambda_i - \Lambda_r v)\ell]}{\Lambda_i - \Lambda_r v}. \qquad (2.48)$$

Further, for the transmitted light at the bottom of a slab ($x = \ell, \alpha = 0$):

$$V^i_r(\ell) = \frac{\exp[-\Lambda_r v \ell] - \exp[-\Lambda_i \ell]}{\Lambda_i - \Lambda_r v}. \qquad (2.49)$$

Summing up, we underline that for calculation of the Stokes vector of the diffused singly scattered light in a given plane-parallel slab we need to find, first of all, the eigenvalues and eigenvectors of the extinction matrix $\hat{\sigma}_{ext}$ in this medium. This can easily be done, using various analytical and numerical approaches. Next we must calculate the traces (see Equations 2.44 and 2.45). And, finally, the Stokes vector is calculated using Equation 2.39.

Let us demonstrate the procedure described for a special case of media with optically active spherical particles. The relation of extinction and scattering matrices of media with such particles to their refractive indices and sizes is well known

(Bohren and Huffman, 1983; Kokhanovsky, 2001). In particular, the extinction matrix in this case simplifies to the following general form:

$$\hat{\sigma}_{ext} = \begin{pmatrix} \varepsilon & 0 & 0 & -b \\ 0 & \varepsilon & a & 0 \\ 0 & -a & \varepsilon & 0 \\ -b & 0 & 0 & \varepsilon \end{pmatrix}, \quad (2.50)$$

where $a = 4\pi N k^{-2} \text{Re}(S_{12}(0))$, $b = 4\pi N k^{-2} \text{Im}(S_{12}(0))$, $\varepsilon = 4\pi N k^{-2} \text{Re}(S_{11}(0))$ as specified above. Then we have performing the diagonalization procedure (see Equation 2.40): $\Lambda_1 = \varepsilon - b, \Lambda_2 = \varepsilon + ia, \Lambda_3 = \varepsilon - ia, \Lambda_4 = \varepsilon + b$ and

$$\hat{G} = \begin{pmatrix} 1 & 0 & 0 & -1 \\ 0 & -i & i & 0 \\ 0 & 1 & 1 & 0 \\ 1 & 0 & 0 & 1 \end{pmatrix}. \quad (2.51)$$

Let us find traces (Equations 2.44 and 2.45) now. First of all we note, that $2*2$ two-dimensional matrices $\hat{D}_1$ and $\hat{D}_2$ have the simplified forms, given by Equation 2.34 in the case under study. The evaluation of traces gives:

$$\Psi_{11}^1 = \Psi_{11}^4 = \Psi_{22}^2 = \Psi_{22}^3 = \Psi_{33}^2 = \Psi_{33}^3 = \Psi_{44}^1 = \Psi_{44}^4 = \Psi_{14}^1 = -\Psi_{14}^4 = \Psi_{41}^1 = -\Psi_{41}^4 = 1 \quad (2.52)$$

and

$$-\Psi_{23}^2 = \Psi_{23}^3 = \Psi_{32}^2 = -\Psi_{32}^3 = i. \quad (2.53)$$

Other values of Ψ_{pk}^r are equal to zero.

Then it follows from Equation 2.43:

$$\hat{F}^i = \frac{1}{2} \begin{pmatrix} V_1^i + V_4^i & 0 & 0 & V_1^i - V_4^i \\ 0 & V_2^i + V_3^i & -i(V_2^i - V_3^i) & 0 \\ 0 & i(V_2^i - V_3^i) & V_2^i + V_3^i & 0 \\ V_1^i - V_4^i & 0 & 0 & V_1^i + V_4^i \end{pmatrix}. \quad (2.54)$$

Thus, the problem of finding the Stokes vector of the diffused light in an asymmetric turbid medium with optically active spherical particles is reduced to simple matrix multiplications (see Equations 2.39 and 2.42).

The assumption of a symmetric medium gives: $\hat{F} = V h_4$, where h_4 is the $4*4$ unity matrix and (see Equation 2.46):

$$V(x) = \frac{\exp[-\Lambda v x] - \exp[-\Lambda x + (1-v)\Lambda a]}{\Lambda(1-v)}, \quad (2.55)$$

where Λ is the extinction coefficient. We accounted for the fact that $\hat{\sigma}_{ext} = \Lambda h_4$ for the symmetric isotropic media. Then we have (see Equation 2.39):

$$\vec{I} = V \hat{\sigma}_{sca}^d \vec{J}, \quad (2.56)$$

which is a familiar result (Hansen and Travis, 1974).

Specifically, we have for the reflected light at the upper boundary ($x = 0, \alpha = \ell$):

$$V^{ref} = \frac{1 - \exp\left[-\Lambda z_0 \left(\frac{1}{\cos \vartheta_0} - \frac{1}{\cos \vartheta}\right)\right]}{(1-v)\Lambda}. \tag{2.57}$$

Also we obtain for the transmitted diffused light at the bottom of a slab ($x = \ell$, $\alpha = 0$):

$$V^{tr} = \frac{\exp\left[-\frac{\Lambda z_0}{\cos \vartheta_0}\right] - \exp\left[-\frac{\Lambda z_0}{\cos \vartheta}\right]}{(1-v)\Lambda}. \tag{2.58}$$

Indices *ref* and *tr* indicate reflected or transmitted light correspondingly.

The angles ϑ and ϑ_0 are counted starting from the positive direction of the axis OZ (downward vertical, see Figure 2.1). Thus, the value of $\mu_0 = \cos \vartheta_0$ is always positive for the radiation entering a light scattering layer from the top. The same applies to the value of $\cos \vartheta$ in the case of transmitted radiation. However, the value of $\cos \vartheta$ is always a negative one for the reflected light. Thus, introducing $\mu = |\cos \vartheta|$, we obtain:

$$V^{ref} = \frac{1 - \exp\left[-\Lambda z_0 \left(\frac{1}{\mu} + \frac{1}{\mu_0}\right)\right]}{(\mu_0 + \mu)\Lambda} \mu_0, \tag{2.59}$$

$$V^{tr} = \frac{\exp\left[-\frac{\Lambda z_0}{\mu_0}\right] - \exp\left[-\frac{\Lambda z_0}{\mu}\right]}{(\mu_0 - \mu)\Lambda} \mu_0. \tag{2.60}$$

Combining Equations 2.56, 2.59, and 2.60, we finally obtain for the Stokes vectors of reflected $\vec{I}^{ref}$ and transmitted $\vec{I}^{tr}$ light:

$$\left.\begin{array}{l} \vec{I}^{ref} = \dfrac{1 - \exp\left[-\Lambda z_0 \left(\dfrac{1}{\mu} + \dfrac{1}{\mu_0}\right)\right]}{(\mu_0 + \mu)\Lambda} \mu_0 \hat{\sigma}^d_{sca} \vec{J}, \\[2ex] \vec{I}^{tr} = \dfrac{\exp\left(-\dfrac{\Lambda z_0}{\mu_0}\right) - \exp\left(-\dfrac{\Lambda z_0}{\mu}\right)}{(\mu_0 - \mu)\Lambda} \mu_0 \hat{\sigma}^d_{sca} \vec{J}. \end{array}\right\} \tag{2.61}$$

Let us introduce the reflection vector function now

$$\vec{\Re} = \frac{\pi \vec{I}^{ref}}{\mu_0 \Im_0}, \tag{2.62}$$

where $\Im_0$ is the incident light flux density on the area perpendicular to the incident beam. Then it follows from Equations 2.61 and 2.62:

$$\vec{\Re} = \frac{\omega_0 \hat{F}(\theta)}{4(\mu_0 + \mu)} \left\{1 - \exp\left[-\left[\frac{1}{\mu} + \frac{1}{\mu_0}\right]\tau\right]\right\} \vec{J}_0, \tag{2.63}$$

where

$$\theta = \arccos\left(-\mu\mu_0 + \sqrt{1-\mu^2}\sqrt{1-\mu_0^2}\cos(\phi-\phi_0)\right),$$

$$\vec{J}_0 = \vec{J}/\Im,$$

$\tau = \Lambda z_0$ is the optical thickness, $\omega_0 = \sigma_{sca}/\Lambda$ is the single scattering albedo and we used the definition of the phase matrix $\hat{F}$ (Equation 2.22). Note that the vector $\vec{J}_0$ has components $(1,0,0,0)$ for incident unpolarized light.

Correspondingly, introducing the transmission vector function

$$\vec{T} = \frac{\pi \vec{I}^{tr}}{\mu_0 \Im}, \qquad (2.64)$$

we obtain:

$$\vec{T} = \frac{\omega_0 \hat{F}(\theta)}{4(\mu_0 - \mu)}\left\{\exp\left[-\frac{\tau}{\mu_0}\right] - \exp\left[-\frac{\tau}{\mu}\right]\right\}\vec{J}_0, \qquad (2.65)$$

where

$$\theta = \arccos\left(\mu\mu_0 + \sqrt{1-\mu^2}\sqrt{1-\mu_0^2}\cos(\phi-\phi_0)\right).$$

Also we have at $\mu = \mu_0$:

$$\vec{T} = \frac{\omega_0 \tau \hat{F}(\theta)}{4\mu_0}\exp\left[-\frac{\tau}{\mu_0}\right]\vec{J}_0.$$

2.3.2 Thick layers

The equations presented in the previous section can be applied only for the case of thin plane-parallel slabs, when multiple light scattering is negligible (e.g., $\tau \leq 0.01$). Another important approximation can be derived for the case of slabs, having large optical thickness τ. Corresponding equations for azimuthally averaged reflection $\hat{R}$ and transmission $\hat{T}$ matrices were obtained by Domke (1978a,b). They have the following forms for symmetric isotropic media:

$$\hat{R}(\mu, \mu_0) = \hat{R}_\infty(\mu, \mu_0) - s\hat{T}(\mu, \mu_0)\exp(-\kappa\tau), \qquad (2.66)$$

$$\hat{T}(\mu, \mu_0) = \frac{w\exp(-\kappa\tau)}{1 - s^2\exp(-2\kappa\tau)}\vec{K}(\mu)\vec{K}^T(\mu_0), \qquad (2.67)$$

where only two-dimensional matrices and vectors are involved. Other components of generally four-dimensional matrices and vectors vanish due the azimuthal averaging. Note that this is also the case for a normal illumination of the scattering layer. Then the azimuth does not enter the theory at all.

Vectors $\vec{\Re}$ and $\vec{T}$ defined in the previous section, are obtained as

$$\vec{\Re} = \hat{R}\vec{J}_0, \qquad \vec{T} = \hat{T}\vec{J}_0. \qquad (2.68)$$

The various parameters and functions in Equations 2.66 and 2.67 are defined as

follows:
$$w = 2\int_0^1 d\mu\mu[\vec{\Theta}^T(\mu)\vec{\Theta}(\mu) - \vec{\Theta}^T(-\mu)\vec{\Theta}(-\mu)], \quad s = 2\int_0^1 d\mu\mu \hat{K}^T(\mu)\vec{\Theta}(-\mu), \quad (2.69)$$

$$\vec{K}(\mu) = \frac{1}{w}\left[\vec{\Theta}(\mu) - 2\int_0^1 d\mu'\mu'\hat{R}_\infty(\mu,\mu')\vec{\Theta}(-\mu')\right]. \quad (2.70)$$

Here
$$\hat{R}_\infty(\mu,\mu') = \frac{1}{2\pi}\int_0^{2\pi}\hat{R}_\infty(\mu,\mu',\psi)d\psi$$

(ψ is the relative azimuth) is the azimuthally averaged reflection matrix $\hat{R}_\infty(\mu,\mu',\psi)$ of a semi-infinite medium with the same optical characteristics as a finite slab under study. The vector

$$\vec{\Theta}(\mu) = \begin{pmatrix} i_1 + i_2 \\ i_1 - i_2 \end{pmatrix}$$

describes the intensity and degree of light polarization in deep layers of a semi-infinite scattering medium (in the so-called asymptotic regime, when the angular distributions of the intensity $i = i_1 + i_2$ and polarization $p_l = (i_2 - i_1)/(i_1 + i_2)$ are symmetrical with respect to the normal to a scattering layer). Note that i exponentially decreases with the depth z ($\sim \exp(-\kappa\sigma_{ext}z)$) in the asymptotic regime. However, angular distributions of the intensity i and polarization p do not change with the depth inside the medium as $\sigma_{ext}z \to \infty$.

We see, therefore, that the azimuthally averaged intensity and polarization characteristics of reflected and transmitted light for optically thick turbid media are determined by the reflection matrix of a semi-infinite layer $\hat{R}_\infty(\mu,\mu_0)$ and the angular distribution of the light intensity and polarization (the vector $\vec{\Theta}(\mu)$) in deep layers of the same medium. This reduction of a problem for a finite optically thick slab to the case of a semi-infinite medium is of a general importance for the radiative transfer theory. Note that the vector $\vec{\Theta}(\mu)$, the characteristic number κ, and the matrix $\hat{R}_\infty(\mu,\mu_0)$ are obtained from solutions of the well-known integral equations (Van de Hulst, 1980; de Rooij, 1985):

$$(1 - \kappa\mu)\vec{\Theta}(\mu) = \frac{\omega_0}{2}\int_{-1}^1 d\mu'\hat{F}^0(\mu,\mu')\vec{\Theta}(\mu')$$

and
$$(\mu + \mu_0)\hat{R}_\infty(\mu,\mu_0) = \frac{\omega_0}{4}\hat{F}^0(-\mu,\mu_0)$$
$$+ \frac{\omega_0}{2}\mu\int_0^1 \hat{R}_\infty(\mu,\mu')\hat{F}^0(\mu',\mu_0)\,d\mu'$$
$$+ \frac{\omega_0}{2}\mu_0\int_0^1 \hat{F}^0(-\mu,-\mu')\hat{R}_\infty(\mu',\mu_0)\,d\mu'$$
$$+ \omega_0\mu\mu_0\int_0^1 d\mu''\int_0^1 d\mu'\hat{R}_\infty(\mu,\mu')\hat{F}^0(\mu',-\mu'')\hat{R}_\infty(\mu'',\mu_0),$$

where

$$\hat{F}^0(\mu,\mu') = \frac{1}{2\pi}\int_0^{2\pi} \hat{F}^0(\mu,\mu',\psi)\,d\psi.$$

Equations 2.66 and 2.67 are valid only for azimuthally averaged matrices. In practice, however, measurements are performed for a fixed azimuth. The transmission matrix is azimuthally independent in the case of optically thick layers. The azimuthal dependence in the reflected light disappears in some specific cases (e.g., for the case of normal illumination of isotropic light scattering plane-parallel slabs).

Equations 2.66 and 2.67 are simplified for non-absorbing media. Then it follows:

$$\hat{R}(\mu,\mu_0) = \hat{R}^0_\infty(\mu,\mu_0) - \hat{T}(\mu,\mu_0), \tag{2.71}$$

$$\hat{T}(\mu,\mu_0) = \frac{4}{3(\tau + 2q_0)(1-g)} \vec{K}_0(\mu)\vec{K}_0^T(\mu_0), \tag{2.72}$$

where

$$q_0 = \frac{2}{1-g}\int_0^1 d\mu\mu^2 \vec{K}_0^T(\mu)\vec{j} \tag{2.73}$$

Here

$$\vec{j} = \begin{pmatrix} 1 \\ 0 \end{pmatrix} \tag{2.74}$$

is the unity vector,

$$g = \frac{1}{4}\int_0^\pi p(\theta)\sin 2\theta\,d\theta \tag{2.75}$$

is the asymmetry parameter, and

$$\vec{K}_0(\mu) = \frac{3}{4}\left[\mu + 2\int_0^1 d\zeta\zeta^2 \hat{R}^0_\infty(\mu,\zeta)\right]\vec{j} \tag{2.76}$$

is the so-called escape function and $\hat{R}^0_\infty(\mu,\mu_0)$ is the azimuthally averaged reflection matrix of a semi-infinite non-absorbing medium. This matrix is completely determined by the phase matrix $\hat{P}$, introduced above. It does not depend on the single scattering albedo and the optical thickness by definition.

These asymptotic equations are simple in form. However, they can be used only if auxiliary functions and parameters are known. Their calculations, however, can be quite a complex procedure.

However, it appears that for weakly absorbing media, when the single scattering albedo $w_0 = \sigma_{sca}/\sigma_{ext}$ is close to one, simplifications are possible. Then it follows (Kokhanovsky, 2001; Kokhanovsky et al., 2003):

$$\hat{R}(\mu,\mu_0) = \hat{R}^0_\infty(\mu,\mu_0)\exp(-y\hat{D}(\mu,\mu_0)) - \hat{T}(\mu,\mu_0)\exp(-x-y), \tag{2.77}$$

$$\hat{T}(\mu,\mu_0) = t\vec{K}_0(\mu)\vec{K}_0^T(\mu_0). \tag{2.78}$$

where

$$x = \kappa\tau, \quad y = 4\sqrt{\frac{1-\omega_0}{3(1-g)}}, \quad \kappa = \sqrt{3(1-\omega_0)(1-g)},$$

$$\hat{D}(\mu,\mu_0) = \hat{R}_\infty^{0^{-1}}(\mu,\mu_0)\vec{K}_0(\mu)\vec{K}_0^T(\mu_0), \quad t = \frac{\sinh y}{\sinh(x+\sigma y)}$$

is the global transmittance of a scattering layer,

$$\sigma = \frac{1}{2}\int_0^1 K_0(\mu)\mu^2 d\mu \approx 1.07,$$

and

$$\hat{R}_\infty^0(\mu,\mu_0) = \frac{1}{2\pi}\int_0^{2\pi} \hat{R}_\infty^0(\mu,\mu_0,\psi)\,d\psi,$$

where $\hat{R}_\infty^0(\mu,\mu_0,\psi)$ is the reflection matrix of a semi-infinite non-absorbing layer with the same phase matrix as an absorbing layer of a finite thickness under study. Note that the global transmittance t is defined by the following equation:

$$t = 4\int_0^1 \mu_0\,d\mu_0 \int_0^1 \mu_{11}^T(\mu,\mu_0)\,d\mu.$$

Also we have for the spherical albedo:

$$r = 4\int_0^1 \mu_0\,d\mu_0 \int_0^1 \mu R_{11}(\mu,\mu_0)\,d\mu.$$

The two-dimensional vector:

$$\vec{K}_0(\mu) = \begin{pmatrix} K_{01}(\mu) \\ K_{02}(\mu) \end{pmatrix}$$

(see Equation 2.76) describes the polarization and intensity of light in the Milne problem for non-absorbing semi-infinite media (Wauben et al., 1994). The components $K_{01}(\mu)$ and $K_{02}(\mu)$ of this vector were calculated by Chandrasekhar (1950) for Rayleigh particles ($g = 0$) and by Wauben (1992) for spherical particles with the refractive index $n = 1.44$ and the gamma particle size distribution (1.5) with $\mu = 11.3$, $a_0 = 0.83$ μm. The wavelength λ was equal to 0.55 μm. Note that the model of spheres with $\mu = 11.3$, $a_0 = 0.83$ μm, and $n = 1.44$ is usually used to characterize particles in clouds on Venus (Hansen and Travis, 1974). It follows that for the effective size a_{ef}, the effective variance Δ_{ef}, and the asymmetry parameter g, respectively, in this case: $a_{ef} = 1.05$ μm, $\Delta_{ef} = 0.07$, $g = 0.718$. The results of numerical calculations are presented in Tables 2.1 and 2.2.

One can see that the component K_{01} only weakly depends on the phase function. The component K_{02} differs considerably for $g = 0$ and $g = 0.718$, however. The value of K_{02} is small in comparison with K_{01} and increases with the observation angle arccos μ, counted from the outer normal to a light scattering layer. The component K_{02} equals zero at the nadir observation ($\vartheta = 0^0$) for arbitrary g. The

Table 2.1. The dependence K_{01} on the cosine of the observation angle μ at $g=0$ and $g=0.718$ (Chandrasekhar, 1950; Wauben, 1992).

μ	0	0.2	0.4	0.6	0.8	1.0
$g=0$	0.414	0.614	0.784	0.948	1.109	1.269
$g=0.718$	0.305	0.594	0.778	0.949	1.114	1.277

Table 2.2. The dependence K_{02} on the cosine of the observation angle μ at $g=0$ and $g=0.718$ (Chandrasekhar, 1950; Wauben, 1992).

μ	0	0.2	0.4	0.6	0.8	1.0
$g=0$	0.04853	0.03774	0.02378	0.01541	0.00757	0
$g=0.718$	0.003623	0.003582	0.002393	0.001359	0.000582	0

degree of linear polarization of transmitted light for incident unpolarized light $p_l = -(T_{21}/T_{11})$ does not depend on the optical thickness. It is equal to $-(K_{02}/K_{01})$. The value of p_l for transmitted light changes from zero at $\mu = 1$ to 0.117 at $\mu = 0$ for Rayleigh particles and from zero at $\mu = 1$ to 0.012 at $\mu = 0$ for particles with $a_{ef} = 1.05\,\mu m$ (see Table 2.3). Thus, the degree of polarization of transmitted light is extremely low as one might expect. This is due to the fact that the polarization of light is mostly due to single scattering of incident light. However, the probability of the existence of singly scattered photons at large values of the optical thickness is low. Note that the ellipticity is equal to zero and $p_l \geq 0$, which means that light is polarized in the plane perpendicular to the meridional plane.

Functions $R^0_{\infty 11}(1,\mu)$, $R^0_{\infty 21}(1,\mu)$, and the degree of polarization

$$p_l(1,\mu) = -\frac{R^0_{\infty 21}(1,\mu)}{R^0_{\infty 11}(1,\mu)} \tag{2.79}$$

for incident unpolarized light are presented in Table 2.4 (for the same case as in the last row of Table 2.3). The ellipticity again is equal to zero. One can see that the degree of polarization of light reflected from a semi-infinite layer with Venusian cloud particles (illuminated along normal by unpolarized solar light) is small.

Table 2.3. The degree of linear polarization $-(K_{02}/K_{01})$ (in percent) obtained from data given in Tables 2.1 and 2.2.

μ	0	0.2	0.4	0.6	0.8	1.0
$g=0$	11.72	6.147	3.033	1.626	0.683	0
$g=0.718$	1.188	0.603	0.306	0.143	0.051	0

Table 2.4. The dependence of functions $R^0_{\infty 11}(1,\mu)$, $R^0_{\infty 21}(1,\mu)$, and $p_l(1,\mu)$ at the nadir illumination on the cosine of the observation angle μ for the model of Venusian particles with $g = 0.718$ (Wauben, 1992).

μ	0	0.2	0.4	0.6	0.8	1.0
$R^0_{\infty 11}(1,\mu)$	0.3929	0.7069	0.8489	0.9541	1.061	1.231
$R^0_{\infty 21}(1,\mu)$	−0.01426	−0.01786	−0.01578	−0.01368	−0.01244	0
$p_l(1,\mu)$	0.036	0.025	0.019	0.014	0.012	0

Table 2.5. The dependence of functions $R^0_{\infty 11}(1,\mu)$, $R^0_{\infty 21}(1,\mu)$, and $p_l(1,\mu)$ at the nadir illumination on the cosine of the observation angle μ for Rayleigh atmosphere (Sobolev, 1956).

μ	0	0.2	0.4	0.6	0.8	1.0
$R^0_{\infty 11}(1,\mu)$	0.6	0.77	0.88	0.98	1.07	1.15
$R^0_{\infty 21}(1,\mu)$	−0.27	−0.24	−0.18	−0.118	−0.064	0
$p_l(1,\mu)$	0.45	0.31	0.2	0.12	0.06	0

Note that it follows: $R^0_{\infty 11}(1,\mu) = R^0_{\infty 11}(\mu,1)$ and $p_l(1,\mu) = p_l(\mu,1)$, but $R^0_{\infty 21}(1,\mu) \neq R^0_{\infty 12}(\mu,1)$ at any μ. The values of $R^0_{\infty 11}(1,\mu)$, $R^0_{\infty 21}(1,\mu)$, and $p_l(1,\mu)$ for Rayleigh particles ($g = 0$) are given in Table 2.5. We see that values of $R^0_{\infty 11}(1,\mu)$ are rather close to each other at $g = 0$ and $g = 0.718$. This is not the case for $R^0_{\infty 21}(1,\mu)$ and $p_l(1,\mu)$.

Formulae 2.77 and 2.78 can be simplified for non-absorbing media ($y = 0$):

$$\hat{R}(\mu,\mu_0) = \hat{R}^0_\infty(\mu,\mu_0) - \hat{T}(\mu,\mu_0), \tag{2.80}$$

$$\hat{T} = t\vec{K}_0(\mu)\vec{K}_0^T(\mu_0), \tag{2.81}$$

where

$$t = \frac{1}{\sigma + \frac{3}{4}\tau(1-g)} \tag{2.82}$$

is the global transmittance.

Let us apply Equation 2.80 to a particular problem, namely, to the derivation of a relation between the spherical albedo r and the degree of polarization of reflected light $p_l(\mu)$ at the illumination along the normal to the scattering layer ($\mu_0 = 1$) by a wide unidirectional unpolarized light beam. The value of $p_l(\mu)$ is given simply by $-R_{21}(1,\mu)/R_{11}(1,\mu)$ in this case. Thus, it follows from Equation 2.80:

$$p_l(\mu) = \frac{p_{l\infty}(\mu)}{1 - (1-r)u(\mu)}, \tag{2.83}$$

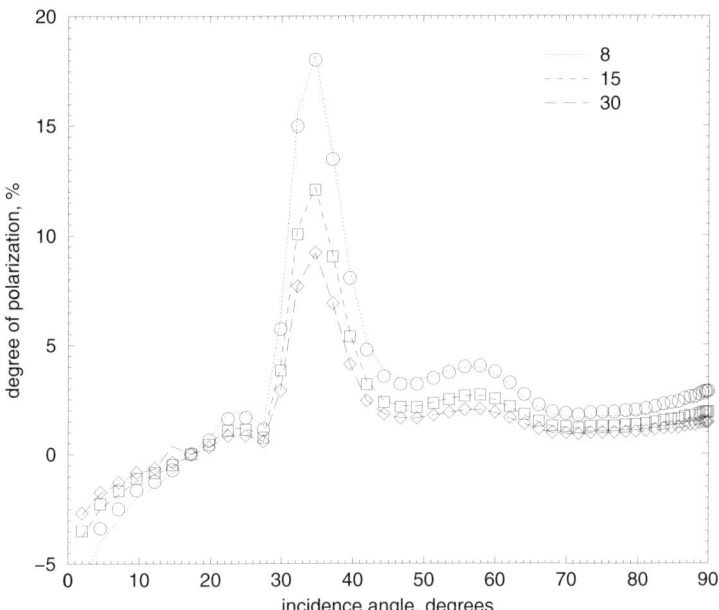

Figure 2.2. The dependence of the degree of polarization of reflected light on the incidence angle at the nadir observation for the particle size distribution (1.5) with $a_0 = 4\,\mu\text{m}$, $\mu = 6$ at the wavelength 443 nm, the refractive index $n = 1.33$, and the cloud optical thickness equal to 8, 15, and 30. Symbols give results of exact calculations. Lines were obtained using Equation 2.83.

where

$$u(\mu) = \frac{K_{01}(1)K_{01}(\mu)}{R^0_{\infty 11}(1,\mu)}, \quad (2.84)$$

$$p_{l\infty}(\mu) = -\frac{R^0_{\infty 21}(1,\mu)}{R^0_{\infty 1}(1,\mu)} \quad (2.85)$$

and we accounted for the equality, $K_{02}(1) = 0$. Also we used the fact that $t = 1 - r$ for non-absorbing media.

Our calculations show that the value of $u(\mu)$ is close to 1 for most observation angles, which implies the inverse proportionality (see Equation 2.83) between the brightness of a turbid medium given by r, and the degree of polarization p_e of reflected light ($rp_l \approx p_{l\infty}$).

This inverse proportionality between the spherical albedo r and the degree of polarization p_l was discovered experimentally by Umow (1905). Equation 2.83 can be considered as a manifestation of this important law, which has important applications in reflectance spectroscopy (Hapke, 1993).

Equation 2.83 remains valid, if the value of μ is substituted by μ_0, but we assume that the observation is at the nadir geometry. The accuracy of Equation 2.83 as the function of the incidence angle is studied in Figure 2.2, where data

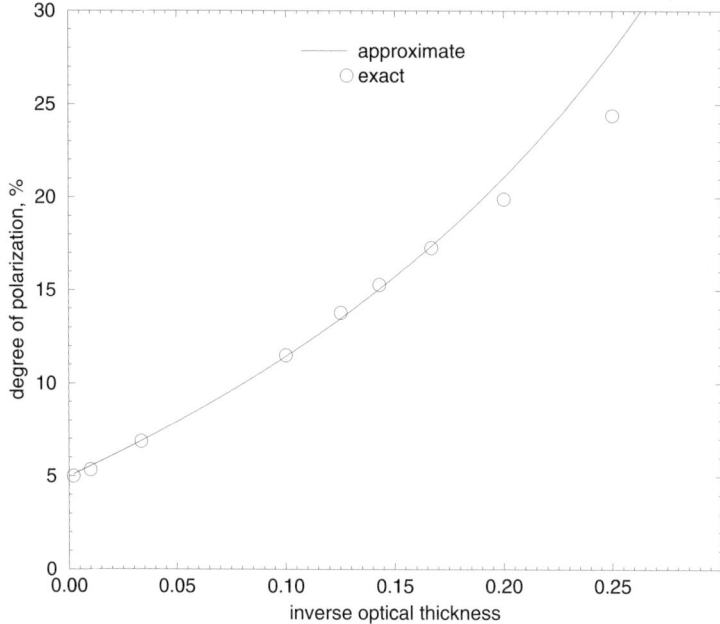

Figure 2.3. The same as in Figure 2.2 but as the function of the inverse optical thickness at the incidence angle 37°.

obtained from this simple formula and results of the numerical solution of the vector radiative transfer equation for cloudy media by the doubling method (van de Hulst, 1980) are presented. We have found that the function $u(\mu)$ varies from 1 to 1.36 for all μ. The variation of the function $u(\mu)$ with the observation angle was neglected. It was assumed that $u = 1.18$ in a first coarse approximation. The value of $p_{l\infty}(\mu)$ has been obtained from the exact radiative transfer calculations for a semi-infinite layer. The gamma PSD was given by Equation (1.5) with the mode radius equal to 4 µm and the half-width parameter equal to 6. The optical thickness of a layer was equal to 8, 15, and 30. Calculations of the phase matrix were performed with the Mie theory at the wavelength 443 nm, where light absorption by water droplets can be neglected. The asymmetry parameter was equal to 0.8541 in the case under investigation.

It follows from Figure 2.2 that Equation 2.83 describes the transformation of the polarization curves due to the change in the optical thickness with a high accuracy at $\tau \geq 8$. As a matter of fact, it can be used down to the optical thickness equal to 5 (see Figure 2.3).

Equation 2.83 is easily generalized to account for the absorption of light in a medium. Namely, it follows:

$$p_l(\mu) = \frac{p^*_{l\infty}(\mu)}{1 - u^*(\mu) t \exp(-x - y)}, \qquad (2.86)$$

where
$$t = \frac{\sinh y}{\sinh(x + \sigma y)} \qquad (2.87)$$
is the global transmittance and
$$u^*(\mu) = \frac{K_{01}(1)K_{01}(\mu)}{R^*_{\infty 11}(1,\mu)}. \qquad (2.88)$$

Values of $p^*_{l\infty}(\mu)$ and $R^*_{\infty 11}(1,\mu)$ represent the degree of polarization and reflection function of a semi-infinite weakly absorbing medium at the nadir illumination. Note that we have approximately (Kokhanovsky, 2001):
$$R^*_{\infty 11}(1,\mu) = R^0_{\infty 11}\exp(-uy), \qquad p^*_{l\infty}(1,\mu) = p^0_{l\infty}\exp(uy).$$

Equation 2.86 can be written in the following form:
$$p_l(\mu) = c(\mu, \tau)p^*_{l\infty}(\mu), \qquad (2.89)$$
where
$$c(\mu, \tau) = \frac{1}{1 - u^*(\mu)t\exp(-x - y)} \qquad (2.90)$$

can be interpreted as the polarization enhancement factor, which is solely due to a finite depth of a turbid layer. It follows for semi-infinite layers that the transmittance t is equal to zero and $c = 1$, as it should be. Also it follows from Equation 2.86 that zeros of polarization curves for semi-infinite and optically thick finite layers almost coincide, which is supported by numerical calculations with the radiative transfer code (see Figure 2.2). This is due to the fact that the function $u^*(\mu)$ only weakly depends on the angle.

Note that multiple light diminishes the polarization of singly scattered light. Thus, the angles, where the polarization is equal to zero for singly scattered light are well-preserved for semi-infinite layers as well.

Formulae presented above could be useful in the solution of inverse problems, particularly those that involve spectro-polarimetric measurements. For this, one should prepare an optically thick layer in the laboratory. In most of the remote sensing applications, however, the investigated medium is not under control, and the numerical solution of Equation 2.21 is not avoidable. Convenient numerical schemes of its solution are given by de Rooij (1985) and de Haan et al. (1987) among others. They will be not discussed here, however.

2.4 FURTHER READING

Apresyan, L. A. and Kravtsov, Y. P. (1983) *Theory of the Radiative Transfer*. Nauka, Moscow.
Bohren, C. F. and Huffman, D. R. (1983) *Absorption and Scattering of Light by Small Particles*. Wiley, New York.
Chandrasekhar, S. (1950) *Radiative Transfer*. Oxford University Press, Oxford, UK.

De Haan, J. F., Bosma, P. B. and Hovenier J. W. (1987) The adding method for multiple scattering calculations of polarized light. *Astron. & Astrophys.*, **183**, 371–391.

De Rooij, W. A. (1985) Reflection and transmission of polarized light by planetary atmospheres. PhD thesis, Amsterdam Free University, Enschede, The Netherlands.

Domke, H. (1978a) Linear Fredholm integral equations for radiative transfer problems in finite plane parallel media. I. Imbedding in an infinite medium. *Astron. Nachr.*, **299**, 87–93.

Domke, H. (1978b) Linear Fredholm integral equations for radiative transfer problems in finite plane parallel media. II. Imbedding in a semi-infinite medium. *Astron. Nachr.*, **299**, 95–102.

Hansen, J. E. and Travis L. D. (1974) Light scattering in planetary atmospheres. *Space Sci. Rev.*, **16**, 527–610.

Hapke, B. (1993) *Theory of the Reflectance and Emmitance Spectroscopy*. Cambridge University Press, Cambridge, UK.

Hovenier, J. W. (1969) Symmetry relationships for scattering of polarized light in a slab of randomly oriented particles. *J. Atmos. Sci.*, **26**, 488–499.

Hovenier, J. W. (1971) Multiple scattering of polarized light in planetary atmospheres. *Astron. & Astrophys.*, **13**, 7–29.

Hovenier, J. W. and de Haan, J. F. (1985) Polarized light in planetary atmospheres for perpendicular directions. *Astron. & Astrophys.*, **146**, 185–191.

Hovenier, J. W., van der Mee, C. V. M. and Domke, H. (in press) *Transfer of Polarized Light in Planetary Atmospheres*. Kluwer, Dordrecht, The Netherlands.

Ishimaru, A. and Yeh, C. W. (1984) Matrix representation of the vector radiative transfer theory for randomly distributed nonspherical particles. *J. Opt. Soc. Am.*, **A1**, 359–364.

Ivanov, A. P. (1969) *Optics of Scattering Media*. Science and Technology Publishing, Minsk.

Jin, Y. Q. (1994) *Electromagnetic Modeling for Quantitative Remote Sensing*. World Scientific, Singapore.

Khvolson, O. D. (1886) Foundations of the mathematical theory of the internal light diffusion. *J. Rus. Phys. – Chem. Society*, **18**, 93–101.

Khvolson, O. D. (1890) Photometric studies of the internal light diffusion. *J. Rus. Phys. – Chem. Society*, **22**, 1–6.

Kokhanovsky, A. A. (2000) The tensor radiative transfer equation. *J. Phys.*, **A33**, 4121–4128.

Kokhanovsky, A. A. (2001) *Light Scattering Media Optics: Problems and Solutions*. Springer-Praxis, Chichester, UK.

Kokhanovsky, A. A., Rozanov, V. V., Zege, E. P., Bovensmann, H., and Burrows, J. P. (2003) A semi-analytical cloud retrieval algorithm using backscattered radiation in 0.4–2.4 µm spectral region. *J. Geophys. Res. D*, **108**, 10.1029/2001JD001543.

Kuga, Y. (1991) Third and fourth-order iterative solutions for the vector radiative transfer equation. *J. Opt. Soc. America*, **8**, 1580–1586.

Kuscher, I. and Ribaric, M. (1959) Matrix formalism in the theory of diffusion of light. *Opt. Acta*, **6**, 42–51.

Mishchenko, M. I. (1990) Multiple scattering of polarized light in anisotropic plane-parallel media. *Transport Theory and Statistical Physics*, **19**, 293–316.

Mishchenko, M. I., Dlugach, J. M., Yanovitskij, E. G., and Zakharova, N. T. (1999) Bidirectional reflectance of flat, optically thick particulate layers: An efficient radiative transfer solution and applications to snow and soil surfaces. *J. Quant. Spectrosc. Radiat. Transfer*, **63**, 409–432.

Mishchenko, M. I. (2002) Vector radiative transfer equation for arbitrarily shaped and arbitrarily oriented particles: A microphysical derivation from statistical electromagnetics. *Appl. Optics*, **41**, 7114–7134.

Mishchenko, M. I., Travis, L. D. and Lacis, A. A. (2002) *Scattering, Absorption, and Emission of Light by Small Particles*. Cambridge University Press, Cambridge, UK.

Nagirner, D. I. (2001) *Lectures on Radiative Transfer Theory*. St Petersburg University Publishing Company, St. Petersburg.

Papanicolaou, G. C. and Burridge, R. (1975) Transport equations for the Stokes parameters from Maxwell equations in random media. *J. Math. Phys.*, **16**, 2074–2082.

Rozenberg, G. V. (1955) Stokes vector-parameter. *Uspekhi Fiz. Nauk*, **56**, 77–110.

Saxon, D. S. (1955) *Lectures on the Scattering of Light*, (Sci.Rep. 9). UCLA Department of Meteorology, Los Angeles.

Sobolev, V. V. (1956) *Radiative Transfer in Stellar and Planetary Atmospheres*. Gostekhteorizdat, Moscow.

Sobolev, V. V. (1972) *Light Scattering in Planetary Atmospheres*, Nauka, Moscow.

Tsang, L., Kong, J. A. and Shin, R. (1985) *Theory of Microwave Remote Sensing*. Wiley, New York.

Tsang, L., Kong, J. A. and Ding, K. H. (2000) *Scattering of Electromagnetic Waves: Theories and Applications*. Wiley, New York.

Tsang, L., Kong, J. A., Ding, K. H. and Ao, C. O. (2001) *Scattering of Electromagnetic Waves: Numerical Simulations*. Wiley, New York.

Tsang, L. and Kong, J. A. (2001) *Scattering of Electromagnetic Waves: Advanced Topics*. Wiley, New York.

Tynes, H. H., Kattawar, G. W., Zege, E. P., Katsev, I. L., Prikhach, A. S. and Chaikovskaya, L. I. (2001) Monte Carlo and multicomponent approximation methods for vector radiative transfer by use of effective Mueller matrix calculations. *Appl. Opt.*, **40**, 400–412.

Umow, N. (1905) Chromatische Depolarisation durch Lichtzerstreung. *Phys. Z.*, **6**, 674–676.

Van de Hulst, H. C. (1980) *Multiple Light Scattering: Tables, Formulas and Applications*. Academic Press, New York.

Van de Hulst, H. C. (1981) *Light Scattering by Small Particles*. Wiley, New York.

Wauben, W. M. F. (1992) Multiple scattering of polarized radiation in planetary atmospheres. PhD thesis, Free University of Amsterdam, Enschede.

Wauben, W. M. F. and Hovenier, J. W. (1992) Polarized radiation in an atmosphere containing randomly-oriented spheroids. *J. Quant. Spectr. Radiat. Transfer*, **47**, 491–504.

Wauben, W. M. F., de Haan, J. F. and Hovenier, J. W. (1993) Low orders of scattering in a plane-parallel homogeneous atmosphere. *Astron. & Astrophys.*, **276**, 589–602.

Wauben, W. M. F., de Haan, J. F. and Hovenier, J. W. (1994) Application of asymptotic expressions for computing the polarized radiation in optically thick planetary atmospheres. *Astron. & Astrophys.*, **281**, 258–268.

Zege, E. P., Kokhanovsky, A. A. and Chaikovskaya, L. I. (1987) Light field in deep layers of an isotropic light scattering medium with soft optically active particles. *Vesti of Belarussian Academy of Sciences, Physics & Mathematics*, **5**, 61–69.

Zege, E. P. and Chaikovskaya, L. I. (2000) Approximate theory of linearly polarized light propagation through a scattering medium. *J. Quant. Spectr. Radiative Transfer*, **66**, 413–435.

3

Local optical characteristics

3.1 GENERAL EQUATIONS

3.1.1 The phase matrix

The solution of Equation 2.21 for the Stokes vector depends on two parameters (ω_0, τ) and the phase matrix $(\hat{P})$.

The parameter τ gives the number of photon free paths $h = \sigma_{ext}^{-1}$ on the thickness of a scattering layer L. Namely, it follows by definition: $\tau = L/h$. If this number is small (typically, less than 0.01), then secondary and multiple light scattering effects can be neglected. On the other hand, for large values of τ the main contribution comes from multiple light scattering. This explains the reason why light transmitted through thick turbid layers has similar angular distributions which are almost independent of the microphysical parameters of the medium. The single light scattering features are almost lost due to the photon random walk process.

The value of ω_0 gives the probability of photon survival in a single scattering event. It is equal to one for non-absorbing media, and zero for absorbing media which do not scatter. The single scattering albedo ω_0 depends mostly on the size and imaginary part of the refractive index of particles. Clearly, the range of applicability of the single scattering approximation grows with probability of photon absorption $\beta = 1 - \omega_0$. Then, multiple light scattering is diminished. Convenient approximations for the extinction coefficient σ_{ext}, the absorption coefficient $\sigma_{abs} = (1-\beta)\sigma_{ext}$, and the asymmetry parameter g of cloudy media are given in Appendix 4.

The task of this chapter is to study the phase matrix $\hat{P}$ for different types of scatterers. The knowledge of this matrix allows a physical problem for a given discrete random medium to be reduced to a mathematical one, related to the solution of Equation 2.21, which can be done using a variety of numerical techniques. Note that vector radiative transfer codes are also available on the Internet (e.g., the library SCATTERLIB at *http://atol.ucsd.edu/~pflatau/scatlib/index.html*).

The process of the interaction of an electromagnetic wave with a particle can be described by different means. The common approach is based on the definition of the interaction matrix $\hat{A}$:

$$\hat{A} = \begin{pmatrix} A_{11} & A_{12} \\ A_{21} & A_{22} \end{pmatrix}, \qquad (3.1)$$

which relates the electric vectors of the electromagnetic wave before ($\vec{E}_0$) and after ($\vec{E}$) a scattering event (see also Equation 1.25). One can also introduce a so-called amplitude scattering matrix $\hat{S}$ using equality, which is valid in the far field zone: $\hat{A} = \Im\hat{S}$. Here $\Im = \exp[-ik(z-r)]/ikr$, and $r \gg k^{-1}$ is the distance from the particle to the observation point, z is the coordinate in the direction of the incoming wave, $k = 2\pi/\lambda$, and λ is the wavelength. It follows that:

$$\vec{E} = \Im\hat{S}\vec{E}_0. \qquad (3.2)$$

Vectors $\vec{E}_0, \vec{E}$ and the matrix $\hat{S}$ are defined in the coordinate system, attached to the scattering plane. This plane holds incident $\vec{k}_0$ and scattered $\vec{k}$ wave vectors ($\vec{k}_0\vec{E}_0 = \vec{k}\vec{E} = 0$). Therefore, components of the vector $\vec{E}_0 = \begin{pmatrix} E_{01} \\ E_{02} \end{pmatrix}$ represent the projection of this vector on the scattering plane (E_{01}) and on the plane perpendicular to the scattering plane (E_{02}). The same holds for the components of the vector $\vec{E} = \begin{pmatrix} E_1 \\ E_2 \end{pmatrix}$.

The main task of the scattering theory is to find elements of the amplitude scattering matrix $\hat{S}$ for particles of different shapes, orientations, internal structures, and chemical compositions. It can be done by solving Maxwell's equations. The shape of the particles is accounted for with the use of boundary conditions for Maxwell's equations for a specific geometry of scatterers. The phase matrix $\hat{P}$ is related to elements of $\hat{S}$ via a simple equation, which follows from Equation 1.35. We would like to identify here not a compact form as given by Equation 1.35, but explicit relationships. For this, we note that the Stokes vectors of scattered $\vec{S}(I, Q, U, V)$ and incident $\vec{S}_0 = (I_0, Q_0, U_0, V_0)$ light are related as follows:

$$\vec{S} = \beta\hat{m}\vec{S}_0, \qquad (3.3)$$

where $\beta = (kr)^{-2}$ and $\hat{m}$ is the $4 \ast 4$ scattering matrix. It is proportional to the Mueller matrix $\hat{M}$, used for the calculation of polarized light transfer in different optical devices. We have, by comparing Equations 1.30 and 3.3, $\hat{M} = \beta\hat{m}$. The elements of the matrix $\hat{m}$ can be derived from the elements of the amplitude scattering matrix, using the theory presented in Section 1.3. We apply another technique here, however.

It follows from Equation 3.2:

$$E_1 = \Im S_{11}E_{01} + \Im S_{12}E_{02}, \qquad (3.4)$$

$$E_2 = \Im S_{21}E_{01} + \Im S_{22}E_{02}. \qquad (3.5)$$

Thus, the expression for the parameter $I = E_1 E_1^* + E_2 E_2^*$ (see Equation 1.9) is

$$I = \beta(|S_{11}|^2 + |S_{21}|^2)|E_{01}|^2 + \beta(|S_{12}|^2 + |S_{22}|^2)|E_{02}|^2$$
$$+ \beta(S_{11} S_{12}^* + S_{21} S_{22}^*) E_{01} E_{02}^* + \beta(S_{11}^* S_{12} + S_{21}^* S_{22}) E_{01}^* E_{02}. \tag{3.6}$$

On the other hand, one can obtain from Equation 3.3:

$$I = \beta(m_{11} I_0 + m_{12} Q_0 + m_{13} U_0 + m_{14} V_0) \tag{3.7}$$

or

$$I = \beta[(m_{11} + m_{12})|E_{01}|^2 + (m_{11} - m_{12})|E_{02}|^2$$
$$+ (m_{13} + m_{14}) E_{01} E_{02}^* + (m_{13} - m_{14}) E_{01}^* E_{02}]. \tag{3.8}$$

It follows from Equations 3.6 and 3.8:

$$\left.\begin{aligned}
m_{11} &= \tfrac{1}{2}(|S_{11}|^2 + |S_{12}|^2 + |S_{21}|^2 + |S_{22}|^2), \\
m_{12} &= \tfrac{1}{2}(|S_{11}|^2 - |S_{12}|^2 + |S_{21}|^2 - |S_{22}|^2), \\
m_{13} &= \mathrm{Re}\,(S_{11} S_{12}^* + S_{22} S_{21}^*), \\
m_{14} &= \mathrm{Im}\,(S_{11} S_{12}^* - S_{22} S_{21}^*).
\end{aligned}\right\} \tag{3.9}$$

Repeating the same procedure for the Stokes parameters Q, U, V, one obtains:

$$\left.\begin{aligned}
m_{21} &= \tfrac{1}{2}(|S_{11}|^2 - |S_{22}|^2 + |S_{12}|^2 - |S_{12}|^2), \\
m_{22} &= \tfrac{1}{2}(|S_{11}|^2 + |S_{22}|^2 - |S_{12}|^2 - |S_{12}|^2), \\
m_{23} &= \mathrm{Re}\,(S_{11} S_{12}^* - S_{22} S_{21}^*), \\
m_{24} &= \mathrm{Im}\,(S_{11} S_{12}^* + S_{22} S_{21}^*), \\
m_{31} &= \mathrm{Re}\,(S_{11} S_{21}^* + S_{22} S_{12}^*), \\
m_{32} &= \mathrm{Re}\,(S_{11} S_{21}^* - S_{22} S_{12}^*), \\
m_{33} &= \mathrm{Re}\,(S_{11}^* S_{22} + S_{12} S_{21}^*), \\
m_{34} &= \mathrm{Im}\,(S_{11} S_{22}^* + S_{21} S_{12}^*), \\
m_{41} &= \mathrm{Im}\,(S_{11}^* S_{21} + S_{12}^* S_{22}), \\
m_{42} &= \mathrm{Im}(S_{11}^* S_{21} - S_{12}^* S_{22}), \\
m_{43} &= \mathrm{Im}\,(S_{22} S_{11}^* - S_{12} S_{21}^*), \\
m_{44} &= \mathrm{Re}\,(S_{22} S_{11}^* - S_{12} S_{21}^*).
\end{aligned}\right\} \tag{3.10}$$

Equations 3.9 and 3.10 allow matrix $\hat{m}$ to be found, if the amplitude scattering matrix $\hat{S}$ is known.

For spherical isotropic particles $S_{12} = S_{21} = 0$ (Van de Hulst, 1981) and therefore,

$$\hat{m} = \begin{pmatrix} \frac{|S_{11}|^2 + |S_{22}|^2}{2} & \frac{|S_{11}|^2 - |S_{22}|^2}{2} & 0 & 0 \\ \frac{|S_{11}|^2 - |S_{22}|^2}{2} & \frac{|S_{11}|^2 + |S_{22}|^2}{2} & 0 & 0 \\ 0 & 0 & \mathrm{Re}\,(S_{11}S_{22}^*) & \mathrm{Im}\,(S_{11}S_{22}^*) \\ 0 & 0 & -\mathrm{Im}\,(S_{11}S_{22}^*) & \mathrm{Re}\,(S_{11}S_{22}^*) \end{pmatrix} \qquad (3.11)$$

The same structure of the matrix $\hat{m}$ holds for isotropic non-spherical particles at random orientation. However, $m_{11} \neq m_{22}$ and $m_{33} \neq m_{44}$ in this case. Also, we have for ensembles of randomly oriented non-spherical particles (Mishchenko et al., 2002):

$$\left. \begin{array}{l} m_{11}(0) = m_{22}(0) = m_{33}(0) = m_{44}(0), \\ m_{12}(0) = m_{34}(0) = m_{12}(\pi) = m_{34}(\pi) = 0, \\ m_{22}(\pi) = -m_{33}(\pi), \end{array} \right\} \qquad (3.12)$$

and

$$m_{44}(\pi) = m_{11}(\pi) - 2m_{22}(\pi) \qquad (3.13)$$

for a special case of rotationally symmetrical non-spherical particles.

Anisotropic or chiral scatterers and oriented systems of nonspherical particles are characterized by Stokes matrices in which all the elements are non-zero (see Equations 3.9 and 3.10).

One can see that the matrix $\hat{m}$ is determined by the elements of the amplitude scattering matrix $\hat{S}$. Thus, the description of disperse media with these two matrices is equivalent, if one is interested in studies of scattered light intensities.

The symbol matrix

$$\hat{\mathbb{Q}} = \begin{pmatrix} ** & -* & /* & \circ * \\ *- & -- & /- & \circ - \\ */ & -/ & // & \circ / \\ *\circ & -\circ & /\circ & \circ\circ \end{pmatrix} \qquad (3.14)$$

represents the input and output optical elements, which are necessary for finding particular matrix elements. Symbols $*$, $-$, $/$, and $\circ$ stand for unpolarized light, horizontally linearly polarized light, linearly polarized light at the azimuth 45°, and left-handed circularly polarized light, respectively. For instance, to obtain the element m_{23} one should measure the intensity of light by illuminating a particle with 45° linearly polarized light while analysing the scattered light with a linear horizontal polarizer (see element $\mathbb{Q}_{23}$ in Equation 3.14). Therefore, Equation 3.14 establishes the physical meaning of all matrix elements.

The scattering properties of a particle can be completely described by measuring all the matrix elements m_{ij} as a function of the scattering angle θ. Knowledge of the matrix $\hat{m}(\theta)$ is important for theoretical studies of polarized radiative transfer in disperse media. In particular, the phase matrix $\hat{P}$ (see Equation 2.22) is proportional to the matrix $\hat{m}$. Namely, we have:

$$\hat{P} = \gamma \hat{m}, \qquad (3.15)$$

where the coefficient of proportionality γ can be found, taking into account the normalization condition:

$$\frac{1}{2}\int_0^\pi P_{11}(\theta)\sin\theta\, d\theta = 1 \qquad (3.16)$$

and the equality (Kerker, 1969):

$$2\pi k^2 \int_0^\pi m_{11}(\theta)\sin\theta\, d\theta = C_{sca}. \qquad (3.17)$$

when C_{sca} is the scattering cross section of a particle. It follows that:

$$\gamma = \frac{4\pi}{k^2 C_{sca}} \qquad (3.18)$$

for monodispersed particles. In the case of polydispersions the phase matrix is obtained using the average values $\bar{m}_{ij}$ and $\bar{C}_{sca}$ (see Equation 2.4):

$$P_{ij}(\theta) = \frac{4\pi \bar{m}_{ij}(\theta)}{k^2 \bar{C}_{sca}},$$

where a line above symbols means the average over the geometrical parameters of the particles (e.g., their sizes). In what follows, we will study the phase matrix and not the amplitude scattering matrix. This is due to the fact that in optical experiments one usually measures intensities or correlation functions and not the fields themselves.

Let us now consider some important relationships between elements of the phase matrix. They should be used to verify the absence of errors in the matrix $\hat{P}$, which can occur due to errors in calculations or measurements of this matrix. The $4*4$ phase matrix is constructed from the $2*2$ complex amplitude scattering matrix (eight real parameters). One can see that nine independent relationships between sixteen elements of the matrix $\hat{P}$ exist (eight real parameters minus the irrelevant phase). They can be presented as follows:

$$(P_{11} \pm P_{22})^2 - (P_{12} + P_{21})^2 - (P_{33} \pm P_{44})^2 - (P_{43} \pm P_{34})^2 = 0 \qquad (3.19)$$

$$(P_{11} \pm P_{21})^2 - (P_{12} \pm P_{22})^2 - (P_{13} \pm P_{23})^2 - (P_{14} \pm P_{24})^2 = 0 \qquad (3.20)$$

$$(P_{11} \pm P_{12})^2 - (P_{21} \pm P_{22})^2 - (P_{31} \pm P_{32})^2 - (P_{41} \pm P_{42})^2 = 0 \qquad (3.21)$$

$$P_{13}P_{14} - P_{23}P_{24} - P_{33}P_{34} - P_{43}P_{44} = 0 \qquad (3.22)$$

$$P_{31}P_{41} - P_{32}P_{42} - P_{33}P_{43} - P_{34}P_{44} = 0 \qquad (3.23)$$

$$P_{14}P_{23} + P_{41}P_{32} - P_{13}P_{24} - P_{42}P_{31} = 0 \qquad (3.24)$$

Each of the relationships (Equations 3.19–3.21) gives two equations, depending on the choice of the upper or lower sign. Note that Equations 3.19–3.24 can be derived assuming various illumination and observation conditions together with the fact that the total degree of polarization is not changed due to the scattering process by a single scatterer.

Equations 3.19–3.24 can be applied only to the case of light scattering by a system of identical scatterers. For polydispersed media, however, they become inequalities (the '=' sign should be changed to '≥'). The same is true for an ensemble of particles differing in shape, orientation, chemical composition, and internal structure.

Another approach to the verification of the phase matrix is based on the calculations of eigenvalues of the coherence matrix (Hovenier, 1986). Eigenvalues should be non-negative. For identical scatters only one eigenvalue differs from zero.

3.1.2 The extinction matrix

The extinction matrix $\hat{\sigma}_{ext}$ (see Equation 2.27) describes the coherent transmission of a light field by a turbid layer and can be found from the following considerations. Let us represent the electromagnetic field $\vec{E}$ in a random medium as a sum of coherent $\vec{E}_c = \langle \vec{E} \rangle$ and incoherent $\vec{E}_i$ ($\langle \vec{E}_i \rangle = 0$) components:

$$\vec{E} = \langle \vec{E} \rangle + \vec{E}_i, \qquad (3.25)$$

where the brackets mean averaging over the statistical ensemble. The coherent field is often called the average field and the incoherent field is called the fluctuation (diffused) field. For the average intensity $\langle I \rangle = \langle \vec{E}\vec{E}^* \rangle$ it follows from Equation 3.25:

$$\langle I \rangle = \langle \vec{E}_c \vec{E}_c^* \rangle + \langle \vec{E}_i \vec{E}_i^* \rangle + \langle \vec{E}_c \vec{E}_i^* \rangle + \langle \vec{E}_i \vec{E}_c^* \rangle. \qquad (3.26)$$

The last two terms are equal to zero (see Equation 3.25) and, therefore, one obtains:

$$\langle I \rangle = I_c + I_i, \qquad (3.27)$$

where

$$I_c = |\langle \vec{E} \rangle|^2, \qquad I_i = \langle |\vec{E}_i|^2 \rangle. \qquad (3.28)$$

The coherent field obeys the following equation (Ishimaru and Yeh, 1984):

$$\frac{d\vec{E}_c}{dL} = \Upsilon \hat{S}(0) \vec{E}_c, \qquad (3.29)$$

where

$$\vec{E}_c = E_{c1}\vec{e}_1 + E_{c2}\vec{e}_2 \qquad (3.30)$$

is the coherent field vector in the plane perpendicular to the direction of propagation $\vec{e}_3 = \vec{e}_2 \times \vec{e}_1$,

$$\Upsilon = -\frac{2\pi}{k^2} N, \qquad (3.31)$$

where N is the number of particles in a unit volume and $\hat{S}(0)$ is the amplitude scattering matrix in the forward direction ($\theta = 0$).

Equation 3.29 can be used to introduce the notion of the effective refractive index M. Let us assume that the matrix $\hat{S}(0)$ is reduced to a scalar value $S(0)$. Then the electric vector of the coherent wave will have the following form after wave propagation through length l in a scattering medium (see Equation 3.29) $\vec{E}_c = \vec{E}_0 \exp(\Upsilon S(0)l)$, where $\vec{E}_0$ is the incident field. On the other hand, it follows that for the electric vector of a transmitted wave after passing though a homogeneous slab without any inclusions, $\vec{E}_c = \vec{E}_0 \exp[-ik(M-1)l]$, where it is assumed that the relative refractive index M of a homogeneous slab, placed between values of $L=0$ and $L=l$, is close to 1. The comparison of these two expressions for $\vec{E}_c$ for the effective refractive index of a scattering slab gives:

$$M = 1 - 2\pi i N k^{-3} S(0).$$

This equation will be used in Chapter 4.

The $4*4$ extinction matrix $\hat{\sigma}_{ext}$ is defined by the following equation (Ishimaru and Yeh, 1984)

$$\frac{d\vec{S}_c}{dL} = -\hat{\sigma}_{ext}\vec{S}_c, \tag{3.32}$$

where $\vec{S}_c(I_c, Q_c, U_c, V_c)$ is the coherent part of the Stokes vector with components:

$$I_c = E_{c1}E_{c1}^* + E_{c2}E_{c2}^*, \tag{3.33}$$

$$Q_c = E_{c1}E_{c1}^* - E_{c2}E_{c2}^*, \tag{3.34}$$

$$U_c = E_{c1}E_{c2}^* + E_{c1}^*E_{c2}, \tag{3.35}$$

$$V_c = i(E_{c1}E_{c2}^* - E_{c1}^*E_{c2}). \tag{3.36}$$

Elements of the matrix $\hat{\sigma}_{ext}$ can be expressed in terms of the elements of the amplitude scattering matrix $\hat{S}(0)$ in the forward direction. Indeed, it follows from Equation 3.29:

$$dE_{c1} = (S_{11}(0)E_{c1} + S_{12}(0)E_{c2})\Upsilon\,dL, \tag{3.37}$$

$$dE_{c2} = (S_{21}(0)E_{c1} + S_{22}(0)E_{c2})\Upsilon\,dL. \tag{3.38}$$

On the other hand, we have (see Equation 3.32):

$$dI_c = (\sigma_{ext}^{11}I_c + \sigma_{ext}^{12}Q_c + \sigma_{ext}^{13}U_c + \sigma_{ext}^{14}V_c)\,dL, \tag{3.39}$$

$$dQ_c = (\sigma_{ext}^{11}I_c + \sigma_{ext}^{12}Q_c + \sigma_{ext}^{13}U_c + \sigma_{ext}^{14}V_c)\,dL, \tag{3.40}$$

$$dU_c = (\sigma_{ext}^{11}I_c + \sigma_{ext}^{12}Q_c + \sigma_{ext}^{13}U_c + \sigma_{ext}^{14}V_c)\,dL, \tag{3.41}$$

$$dV_c = (\sigma_{ext}^{11}I_c + \sigma_{ext}^{12}Q_c + \sigma_{ext}^{13}U_c + \sigma_{ext}^{14}V_c)\,dL. \tag{3.42}$$

Substituting Equations 3.33–3.38 into Equations 3.39–3.42, we obtain:

$$\hat{\sigma}_{ext} = \frac{2\pi N}{k^2} \text{Re} \begin{pmatrix} S_{11}(0)+S_{22}(0) & S_{11}(0)-S_{22}(0) & S_{12}(0)+S_{21}(0) & i(S_{21}(0)-S_{12}(0)) \\ S_{11}(0)-S_{22}(0) & S_{11}(0)+S_{22}(0) & S_{12}(0)-S_{21}(0) & i(S_{12}(0)+S_{21}(0)) \\ S_{12}(0)+S_{21}(0) & S_{21}(0)-S_{12}(0) & S_{11}(0)+S_{22}(0) & i(S_{22}(0)-S_{11}(0)) \\ i(S_{21}(0)-S_{12}(0)) & -i(S_{12}(0)+S_{21}(0)) & -i(S_{22}(0)-S_{11}(0)) & S_{11}(0)+S_{22}(0) \end{pmatrix}$$

(3.43)

This important formula gives the relationship between the extinction matrix and the amplitude scattering matrix elements in the forward scattering direction. We see, therefore, that the knowledge of the $2 * 2$ matrix $\hat{S}$ is sufficient to calculate all the elements of the $4 * 4$ matrices in the radiative transfer equation (see, e.g., Equation 2.17).

The elements of the matrix $\hat{\sigma}_{ext}$ obey the following relationships (see Equation 3.43):

$$\sigma_{ext}^{11} = \sigma_{ext}^{22} = \sigma_{ext}^{33} = \sigma_{ext}^{44}, \qquad (3.44)$$

$$\sigma_{ext}^{12} = \sigma_{ext}^{21}, \qquad \sigma_{ext}^{13} = \sigma_{ext}^{31}, \qquad \sigma_{ext}^{14} = \sigma_{ext}^{41}, \qquad (3.45)$$

$$\sigma_{ext}^{23} = -\sigma_{ext}^{32}, \qquad \sigma_{ext}^{24} = -\sigma_{ext}^{24}, \qquad \sigma_{ext}^{34} = -\sigma_{ext}^{43}. \qquad (3.46)$$

All the elements of the extinction matrix are expressed through combinations: $\pm \text{Re}(S_{11}(0) \pm S_{22}(0))$ and $\pm \text{Re}(S_{12}(0) \pm S_{21}(0))$. This matrix is simplified for some specific cases. For instance, for isotropic spherical particles:

$$S_{11}(0) = S_{22}(0), \qquad S_{12}(0) = S_{21}(0) = 0 \qquad (3.47)$$

and the matrix $\hat{\sigma}_{ext}$ reduces to a scalar value:

$$\sigma_{ext} = \frac{4\pi N}{k^2} \text{Re}(S_{11}(0)). \qquad (3.48)$$

Therefore, effects of dichroism and birefringence do not take place in this case. The following relations hold for chiral spheres (Bohren and Huffman, 1983):

$$S_{11}(0) = S_{22}(0), \qquad S_{12}(0) = -S_{21}(0). \qquad (3.49)$$

Then Equation 3.43 is simplified (see also Equation 2.50):

$$\hat{\sigma}_{ext} = \varepsilon \begin{pmatrix} 1 & 0 & 0 & \beta \\ 0 & 1 & \alpha & 0 \\ 0 & -\alpha & 1 & 0 \\ \beta & 0 & 0 & 1 \end{pmatrix}, \qquad (3.50)$$

$$\alpha = \frac{\text{Re}[S_{12}(0)]}{\text{Re}[S_{11}(0)]}, \qquad \beta = -\frac{\text{Im}[S_{12}(0)]}{\text{Re}[S_{11}(0)]}, \qquad \varepsilon \frac{4\pi N}{K^2} \text{Re}(S_{11}(0)). \qquad (3.51)$$

Media having the extinction matrix given by Equation 3.50 rotate the plane of polarization of incident linearly polarized light beams. They are characterized also by dichroism (different extinction of left- and right-hand polarized waves).

One can introduce the absorption matrix $\hat{\sigma}_{abs}$ with the following relation:

$$\hat{\sigma}_{abs} = \hat{\sigma}_{ext} - \hat{\sigma}^t_{sca}, \qquad (3.52)$$

where the $\hat{\sigma}^t_{sca}$ is the total scattering matrix:

$$\hat{\sigma}^t_{sca} = \int_\Omega \hat{\sigma}_{sca}(\Omega) \, d\Omega \qquad (3.53)$$

and

$$\hat{\sigma}_{sca}(\Omega) = \frac{N}{k^2} \hat{m} \qquad (3.54)$$

for monodispersed media.

In conclusion we underline once more that the main task of the scattering theory is the calculation of the amplitude scattering matrix $\hat{S}$. Local optical properties of particles (including absorption characteristics) can be obtained from this matrix. The scattering matrix $\hat{S}$ is used to find the phase and extinction matrices, which provide the complete set of parameters for the radiative transfer equations.

3.1.3 The classification of disperse media

Generally speaking, all disperse media can be separated into two broad classes, namely isotropic and anisotropic ones. The physical properties of isotropic media are the same in all directions, which is not the case for anisotropic media. It is evident that the distribution of randomly oriented identical non-spherical particles is an isotropic medium. The same particles, however, can be oriented by some physical forces (e.g., magnetic or gravitational ones), leading to different physical characteristics of media in planes parallel and perpendicular to the acting force (Khlebtsov et al., 1991, 1999, 2002; Lacoste et al., 1998; Lacoste, 1999). The orientation transforms the isotropic medium to an anisotropic one. In turn, both isotropic and anisotropic media can be separated into two categories as well. Namely, one can introduce symmetrical and asymmetrical media. The centre of any large spherical volume is a centre of symmetry for symmetrical media. Any plane through this centre is a plane of symmetry. These conditions are not fulfilled in asymmetrical media. Thus, the isotropy of asymmetrical media may be only statistical, as the result of an isotropic distribution of small anisotropic elements (Perrin, 1942). The random distribution of left-handed springs can be an example of such an asymmetrical isotropic medium. Note that most biological media are characterized by asymmetry. Polarization characteristics of light in symmetrical and asymmetrical media differ considerably. One can find general forms of the Stokes matrices for symmetrical and asymmetrical isotropic media in Table 3.1 where results presented were obtained by Perrin (1942) based on the reciprocity principle. This principle states that if two incident polarized beams have equal intensities, the inverse emerging beams of the same polarization, which are associated with them, also have equal intensities. Symmetrical media in Table 3.1 include the important cases of spherical polydispersions (water clouds, sulphate aerosols, fogs, and randomly oriented non-spherical and layered spherical particles). Asymmetrical

Table 3.1. General forms of the normalized phase matrices $\hat{p} = \hat{P}/P_{11}$ for isotropic symmetrical and asymmetrical media (Perrin, 1942) in different scattering directions.

Scattering direction	Isotropic symmetrical media	Isotropic asymmetrical media
Side scattering	$\begin{pmatrix} 1 & p_{12} & 0 & 0 \\ p_{12} & p_{22} & 0 & 0 \\ 0 & 0 & p_{33} & p_{34} \\ 0 & 0 & -p_{34} & p_{44} \end{pmatrix}$	$\begin{pmatrix} 1 & p_{12} & p_{13} & p_{14} \\ p_{12} & p_{22} & p_{23} & p_{24} \\ -p_{13} & -p_{23} & p_{33} & p_{34} \\ p_{14} & p_{24} & -p_{34} & p_{44} \end{pmatrix}$
Forward scattering	$\begin{pmatrix} 1 & 0 & 0 & 0 \\ 0 & p_{22}(0) & 0 & 0 \\ 0 & 0 & p_{22}(0) & 0 \\ 0 & 0 & 0 & p_{44}(0) \end{pmatrix}$	$\begin{pmatrix} 1 & 0 & 0 & p_{14}(0) \\ 0 & p_{22}(0) & p_{23}(0) & 0 \\ 0 & -p_{23}(0) & p_{22}(0) & 0 \\ p_{14}(0) & 0 & 0 & p_{44}(0) \end{pmatrix}$
Backward scattering	$\begin{pmatrix} 1 & 0 & 0 & 0 \\ 0 & p_{22}(\pi) & 0 & 0 \\ 0 & 0 & -p_{22}(\pi) & 0 \\ 0 & 0 & 0 & p_{44}(\pi) \end{pmatrix}$	$\begin{pmatrix} 1 & 0 & 0 & p_{14}(\pi) \\ 0 & p_{22}(\pi) & 0 & 0 \\ 0 & 0 & -p_{22}(\pi) & 0 \\ p_{14}(\pi) & 0 & 0 & p_{44}(\pi) \end{pmatrix}$

Table 3.2. As Table 3.1 but for spherical particles.

Scattering direction	Isotropic symmetrical media	Isotropic asymmetrical media
Side scattering	$\begin{pmatrix} 1 & p_{12} & 0 & 0 \\ p_{12} & 1 & 0 & 0 \\ 0 & 0 & p_{33} & p_{34} \\ 0 & 0 & -p_{34} & p_{33} \end{pmatrix}$	$\begin{pmatrix} 1 & p_{12} & p_{13} & p_{14} \\ p_{12} & p_{22} & p_{23} & p_{24} \\ -p_{13} & -p_{23} & p_{33} & p_{34} \\ p_{14} & p_{24} & -p_{34} & p_{44} \end{pmatrix}$
Forward scattering	$\begin{pmatrix} 1 & 0 & 0 & 0 \\ 0 & 1 & 0 & 0 \\ 0 & 0 & 1 & 0 \\ 0 & 0 & 0 & 1 \end{pmatrix}$	$\begin{pmatrix} 1 & 0 & 0 & p_{14}(0) \\ 0 & 1 & p_{23}(0) & 0 \\ 0 & -p_{23}(0) & 1 & 0 \\ p_{14}(0) & 0 & 0 & 1 \end{pmatrix}$
Backward scattering	$\begin{pmatrix} 1 & 0 & 0 & 0 \\ 0 & 1 & 0 & 0 \\ 0 & 0 & -1 & 0 \\ 0 & 0 & 0 & -1 \end{pmatrix}$	$\begin{pmatrix} 1 & 0 & 0 & 0 \\ 0 & 1 & 0 & 0 \\ 0 & 0 & -1 & 0 \\ 0 & 0 & 0 & -1 \end{pmatrix}$

media in Table 3.1 differ from symmetrical media due to the intrinsic asymmetry of particles. This intrinsic asymmetry of scatterers is common for bio-particles and often manifests itself in the rotatory dispersion and circular dichroism phenomena. Table 3.1 simplifies in the case of spherical particles (see Table 3.2).

3.2 SPHERICAL PARTICLES

3.2.1 Mie theory

Let us consider now the results of numerical calculations of the phase matrices for various types of scattering particles. Clearly, these matrices depend on the size, shape, internal structure, and refractive index of particles. We start with the simplest case, namely the case of spherical isotropic homogeneous particles. This case has a lot of practical applications (e.g., in cloud optics and the paint industry). This is why we consider light scattering by spheres in considerable detail here.

For a single sphere, we have (Bohren and Huffman, 1983):

$$\hat{P} = \gamma \begin{pmatrix} \frac{|S_{11}|^2 + |S_{22}|^2}{2} & \frac{|S_{11}|^2 - |S_{22}|^2}{2} & 0 & 0 \\ \frac{|S_{11}|^2 - |S_{22}|^2}{2} & \frac{|S_{11}|^2 + |S_{22}|^2}{2} & 0 & 0 \\ 0 & 0 & \operatorname{Re}(S_{11}S_{22}^*) & \operatorname{Im}(S_{11}S_{22}^*) \\ 0 & 0 & -\operatorname{Im}(S_{11}S_{22}^*) & \operatorname{Re}(S_{11}S_{22}^*) \end{pmatrix}. \quad (3.55)$$

where $\gamma = 4\pi/k^2 \sigma_{sca}$. The elements of the amplitude scattering matrix $S_{11}(\theta)$ and $S_{22}(\theta)$ are given by:

$$S_{11}(\theta) = \sum_{n=1}^{\infty} \frac{2n+1}{n(n+1)} \{a_n \tau_n(\cos\theta) + b_n \pi_n(\cos\theta)\}, \quad (3.56)$$

$$S_{22}(\theta) = \sum_{n=1}^{\infty} \frac{2n+1}{n(n+1)} \{a_n \pi_n(\cos\theta) + b_n \tau_n(\cos\theta)\}, \quad (3.57)$$

where

$$a_n = \frac{\psi_n'(y)\psi_n(x) - m\psi_n(y)\psi_n'(x)}{\psi_n'(y)\xi_n(x) - m\psi_n(y)\xi_n'(x)}, \quad b_n = \frac{m\psi_n'(y)\psi_n(x) - \psi_n(y)\psi_n'(x)}{m\psi_n'(y)\xi_n(x) - \psi_n(y)\xi_n'(x)}, \quad (3.58)$$

$m = n - i\chi$ is the relative refractive index of a particle ($m = m_p/m_h$, m_p and m_h are the refractive indices of a particle and the host medium respectively), $y = mx$, $x = (2\pi a/\lambda)$ is the size parameter, a is the radius of a sphere, λ is the incident wavelength in the non-absorbing host medium, $\psi_n(x) = \sqrt{(\pi x/2)}J_{n+1/2}(x)$, $\xi_n(x) = \sqrt{(\pi x/2)}H^{(2)}_{n+1/2}(x)$, $J_{n+1/2}$ and $H^{(2)}_{n+1/2}$ are Bessel and Hankel functions. The angular functions $\pi_n(\cos\theta)$, $\tau_n(\cos\theta)$ are determined by the following equations:

$$\pi_n(\cos\theta) = \frac{P_n^{(1)}(\cos\theta)}{\sin\theta}, \quad \tau_n(\cos\theta) = \frac{dP_n^{(1)}(\cos\theta)}{d\theta}, \quad (3.59)$$

where $P_n^{(1)}(\cos\theta)$ is the associated Legendre polynomial and θ is the scattering angle.

Expressions for the values of the extinction efficiency $Q_{ext} = C_{ext}/\pi a^2$, scattering efficiency $Q_{sca} = C_{sca}/\pi a^2$, the phase function $p(\theta)$ at the natural light illumination, and the asymmetry parameter $g = \frac{1}{2}\int_0^\pi p(\theta)\sin\theta\cos\theta\, d\theta$ are presented in Table 3.3.

Table 3.3. Light scattering characteristics for isotropic homogeneous spheres.

Value	Formula
Q_{ext}	$\dfrac{2}{x^2}\sum_{n=1}^{\infty}(2n+1)\operatorname{Re}(a_n+b_n)$
Q_{sca}	$\dfrac{2}{x^2}\sum_{n=1}^{\infty}(2n+1)[|a_n|^2+|b_n|^2]$
$p(\theta)$	$\dfrac{2\pi(i_1+i_2)}{k^2 C_{sca}}, \quad i_1=|S_1|^2, \quad |i_2|=|S_2|^2, \quad C_{sca}=\pi a^2 Q_{sca}$
g	$\dfrac{4}{x^2 Q_{sca}}\sum_{n=1}^{\infty}\left[\dfrac{n(n+2)}{n+1}\operatorname{Re}(a_n a_{n+1}^* + b_n b_{n+1}^*) + \dfrac{2n+1}{n(n+1)}\operatorname{Re}(a_n b_n^*)\right]$

Note that for polydispersed media:

$$\sigma_{ext} = N\int_0^{\infty}\pi a^2 Q_{ext} f(a)\,da, \qquad (3.60)$$

$$\sigma_{sca} = N\int_0^{\infty}\pi a^2 Q_{sca} f(a)\,da, \qquad (3.61)$$

$$p(\theta) = \dfrac{2\pi N\int_0^{\infty}(i_1+i_2)f(a)\,da}{k^2 \sigma_{sca}}, \qquad g = \dfrac{\int_0^{\infty}a^2 g Q_{sca} f(a)\,da}{\int_0^{\infty}a^2 Q_{sca} f(a)\,da}. \qquad (3.62)$$

The implementation of the equations presented here in a numerical code is straightforward. Details are given by Bohren and Huffman (1983), who also published their Mie code.

3.2.2 Optical properties of spherical polydispersions

Let us now study the influence of the size of spheres and their refractive index on the phase matrix elements using the equations given above. For this we use the normalized phase matrix:

$$\hat{p}(\theta) = \dfrac{\hat{P}(\theta)}{P_{11}(\theta)} \qquad (3.63)$$

with the element $P_{11} = 1$. All other elements of this matrix are in the range $[-1, 1]$. This simplifies the analysis. We also present separately the phase function $p(\theta) \equiv P_{11}(\theta)$, which plays an important role in scalar radiative transport. As was stated in Chapter 2, $p(\theta)(d\Omega/4\pi)$ gives the probability of photon scattering in the solid angle $d\Omega$ around the direction, specified by the scattering angle θ.

It follows for spherical particles that:

$$\hat{p} = \begin{pmatrix} 1 & b & 0 & 0 \\ b & 1 & 0 & 0 \\ 0 & 0 & a & c \\ 0 & 0 & -c & a \end{pmatrix}.$$

The physical sense of the functions $a = p_{44}$, $b = p_{12}$, and $c = p_{34}$ can be derived from the following experiment. Let us illuminate a spherical particle by a light beam with the Stokes vector-parameter $\vec{S}_0(I_0, Q_0, U_0, V_0)$. Then we have for the scattered light Stokes vector-parameter $\vec{S}(I, Q, U, V)$, taking into account Equations 3.3 and 3.15:

$$I = A(P_{11}I_0 + P_{12}Q_0), \tag{3.64}$$

$$Q = A(P_{12}I_0 + P_{22}Q_0), \tag{3.65}$$

$$U = A(P_{33}U_0 + P_{34}V_0), \tag{3.66}$$

$$V = A(-P_{34}U_0 + P_{44}V_0). \tag{3.67}$$

Note that $A = \beta/\gamma = C_{sca}/4\pi r^2$ does not depend on the scattering angle. Let us define the effective solid angle associated with scattering:

$$\Omega_{ef} = \frac{C_{sca}}{r^2}.$$

Then we have:

$$A = \frac{\Omega_{ef}}{4\pi},$$

Therefore the constant A can be interpreted as a relative effective solid angle.

We have for spheres: $P_{11} = P_{22}$ and $P_{33} = P_{44}$. It should be emphasized that the parameters (I, Q) transform independently from the parameters (U, V) in the case studied. In particular, it means that the scattering of incident unpolarized light $(Q_0 = U_0 = V_0)$ does not produce circularly polarized light and the degree of circular polarization $p_c = V/I$ is equal to zero in this case. For incident left-handed circularly polarized light $(I_0 = V_0 = 1, U_0 = Q_0 = 0)$ we have, however:

$$I = AP_{11}, \quad Q = AP_{12}, \quad U = Ap_{34}, \quad V = AP_{44}. \tag{3.68}$$

Then the degree of circular polarization of incident light $(V_0/I_0 = 1)$ is reduced due to a scattering process and becomes equal to P_{44}/P_{11} or p_{44}. This gives a physical meaning of the element p_{44} of the normalized phase matrix $\hat{p}$.

In the case of the illumination of a particle by horizontally polarized light $(I_0 = -U_0 = 1, Q_0 = V_0 = 0)$ we have:

$$I = AP_{11}, \quad Q = AP_{12}, \quad U = -AP_{33}, \quad V = AP_{34}. \tag{3.69}$$

The degree of circular polarization of scattered light will be equal to P_{34}/P_{11} or p_{34}. Thus, the element p_{34} shows the efficiency of the linear to circular polarization mode conversion due to a scattering event. Note that $p_{34} \equiv 0$ for Rayleigh particles

$(a/\lambda \to 0)$. It means that Rayleigh particles cannot produce circular polarization for incident unpolarized or linearly polarized incident beams.

Using both Equation 3.68 and 3.69 we obtain the degree of polarization q (see Equation 1.22):

$$q = \sqrt{p_{12}^2 + p_{34}^2 + p_{44}^2}.$$

Note that we have for monodispersed spheres (see Equations 3.56 and 3.57):

$$|S_{11}S_{22}^*|^2 = 4|S_{11}|^2|S_{22}|^2,$$

which gives $q = 1$.

Finally, we have for incident unpolarized light ($Q_0 = U_0 = V_0 = 0, I_0 = 1$):

$$I = AP_{11}, \qquad Q = AP_{12}, \qquad U = 0, \qquad V = 0. \tag{3.70}$$

This means that the degree of linear polarization of scattered light $p_l = -Q/I$ is equal to $-P_{12}/P_{11}$ or $-p_{12}$. This concludes the description of physical meaning of the matrix elements p_{12}, p_{34} and p_{44}.

Note that for spherical Rayleigh particles we have (Bohren and Huffman, 1983):

$$p_{12} = -\frac{\sin^2 \theta}{1 + \cos^2 \theta}, \qquad p_{22} = 1, \qquad p_{33} = p_{44} = \frac{2 \cos \theta}{1 + \cos^2 \theta}, \qquad p_{34} = 0. \tag{3.71}$$

Also, the phase function

$$p(\theta) = \tfrac{3}{4}(1 + \cos^2 \theta) \tag{3.72}$$

in this case. We see that Rayleigh particles strongly polarize incident unpolarized light at angles close to 90°. One can obtain for the degree of linear polarization in this case:

$$p_l \equiv -p_{12} = \frac{\sin^2 \theta}{1 - \cos^2 \theta}, \tag{3.73}$$

which is positive for all scattering angles θ.

The influence of the effective radius of particles, the coefficient of variance of the particle size distribution and the complex refractive index of particles on the phase function $p(\theta)$, the degree of linear polarization ($-p_{12}$), and degrees of circular polarization (p_{34}, p_{44}) are presented in Figures 3.1–3.4. Calculations have been performed, using the Mie theory for spherical polydispersions, characterized by the gamma particle size distribution (PSD) (see Equation 1.5). We also show in Figures 3.1–3.4 the degree of polarization $q = \sqrt{p_{12}^2 + p_{33}^2 + p_{34}^2}$ and the scattered light beam entropy (Brosseau and Bicout, 1994)

$$s = -\ln[0.5(1+q)^{(1+q)/2}(1-q)^{(1-q)/2}]. \tag{3.74}$$

It follows from Equation 3.74 that $s_{max} = s(0) = \ln 2$ at $q = 0$. The entropy s, therefore, is in the range $[0, \ln 2]$. It increases with decreasing q. The entropy production in light scattering processes is described in detail by Brosseau and Bicout (1994). Note that for spherical Rayleigh scatterers and monodispersed spheres, $q = 1, s = 0$. For spherical polydispersions, however, s and $1 - q$ may not vanish.

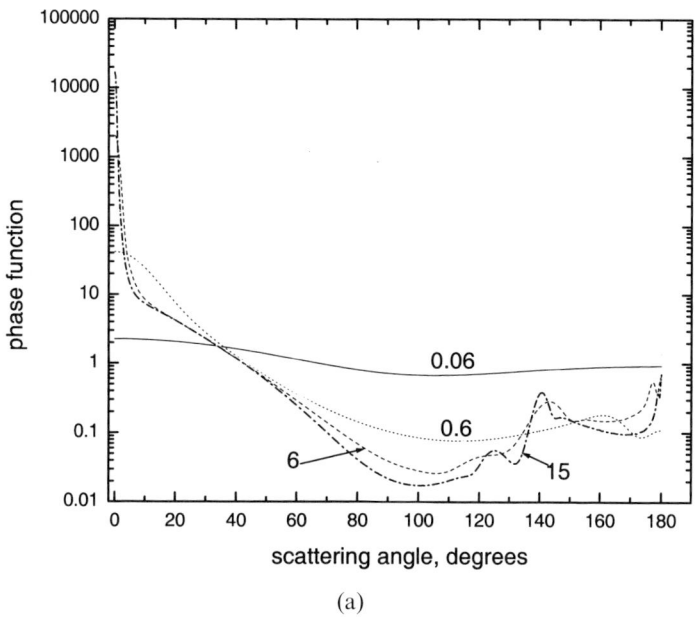

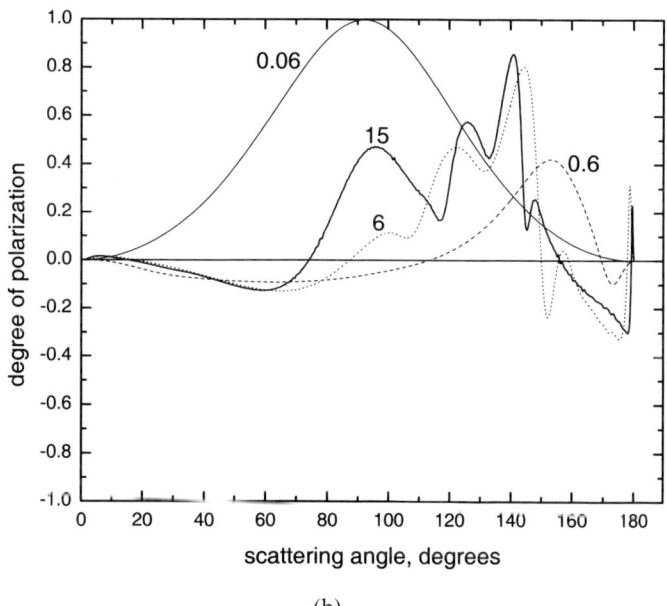

Figure 3.1. (a) The dependence of the phase function of spherical polydispersions with the PSD (1.5) at $\mu = 6$ on the effective radius of droplets $a_{ef} = 0.06\,\mu m$, $0.6\,\mu m$, $6\,\mu m$, and $15\,\mu m$. It is assumed that the wavelength $\lambda = 0.55\,\mu m$ and the refractive index, $m = 1.333$; (b) the same as in (a) but for the degree of polarization p_l.

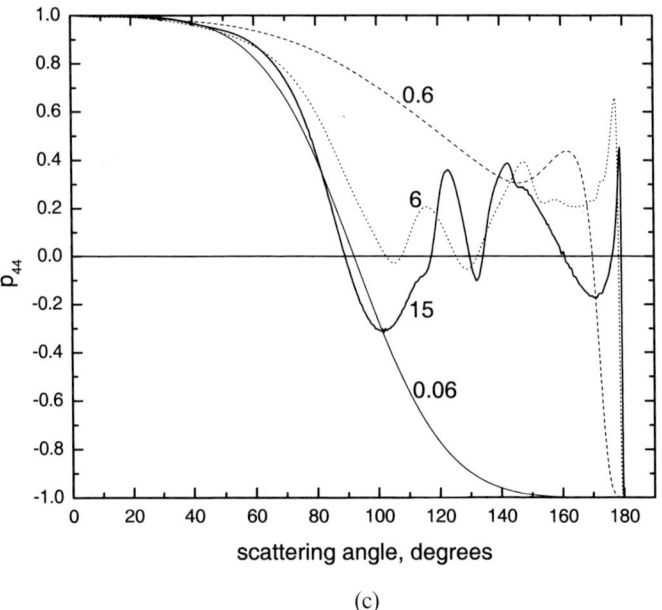

(c)

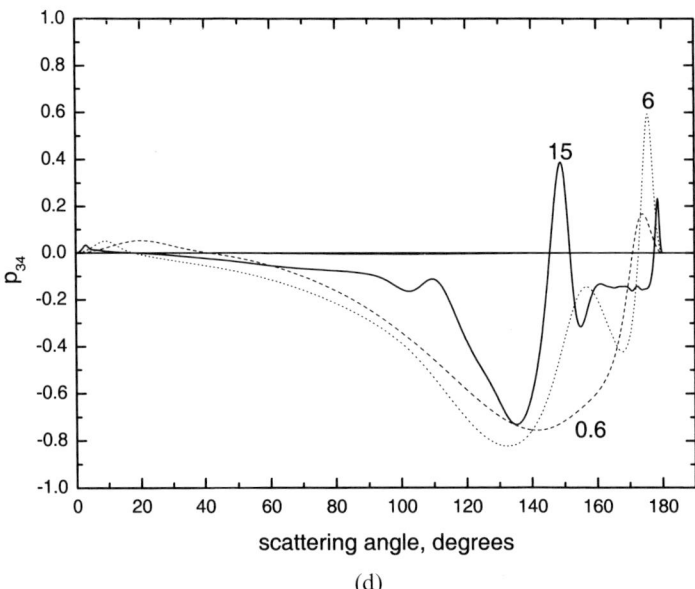

(d)

Figure 3.1 (*cont.*). (c) The same as in (a) but for p_{44}; (d) the same as in (a) but for p_{34}.

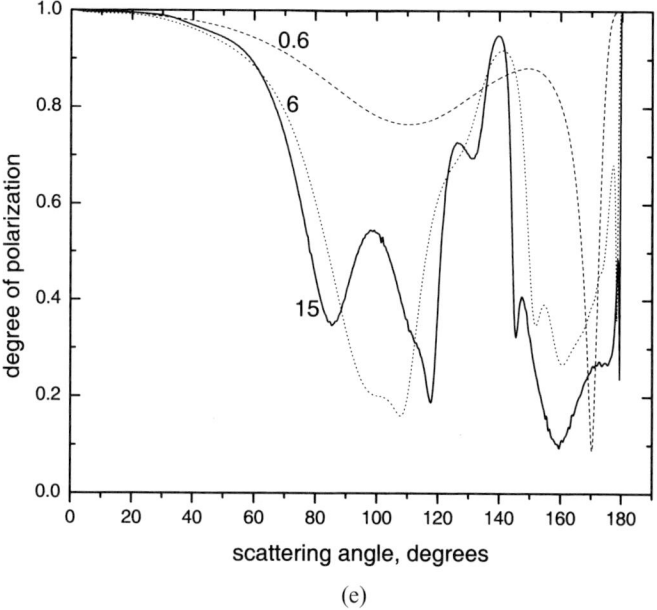

(e)

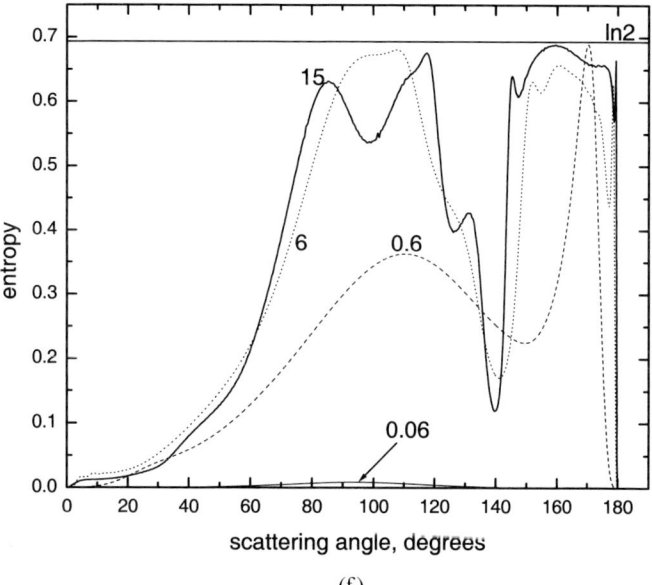

(f)

Figure 3.1 (*cont.*). (e) The same as in (a) but for q; (f) the same as in (a) but for s.

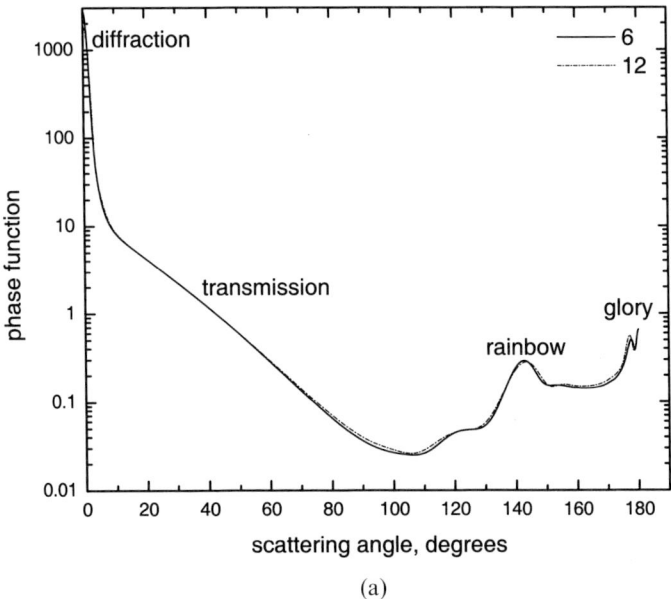

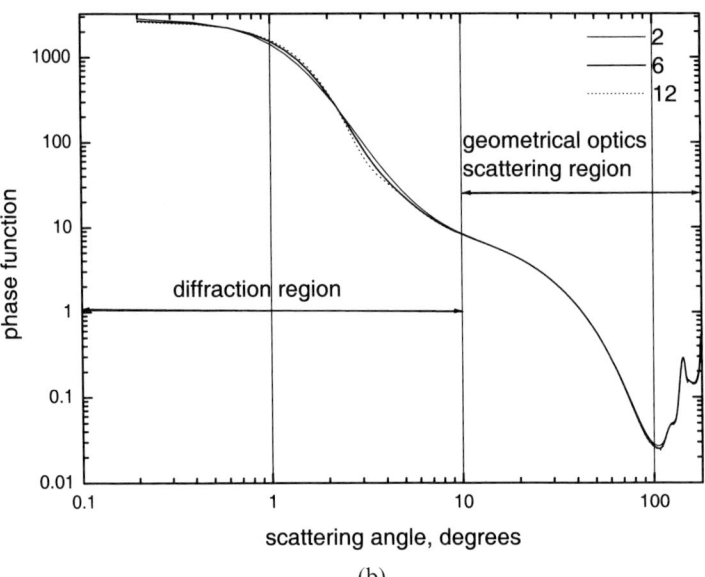

Figure 3.2. (a) The phase function of spherical polydispersions with the PSD (1.5) at $a_{ef} = 6\,\mu m$ and $\mu = 6$ and 12. It is assumed that $\lambda = 0.55\,\mu m$ and $m = 1.333$; (b) the same as in (a) but in a different scale. Also data for $\mu = 2$ are given. Diffraction and geometrical scattering regions are shown.

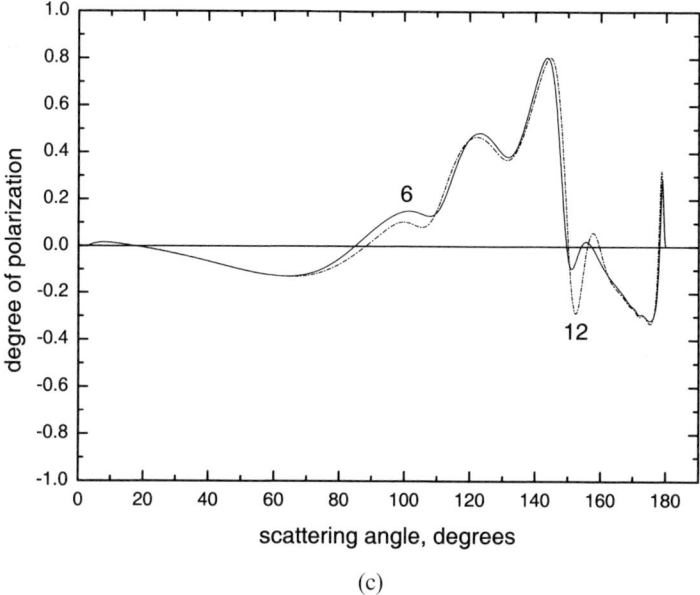

(c)

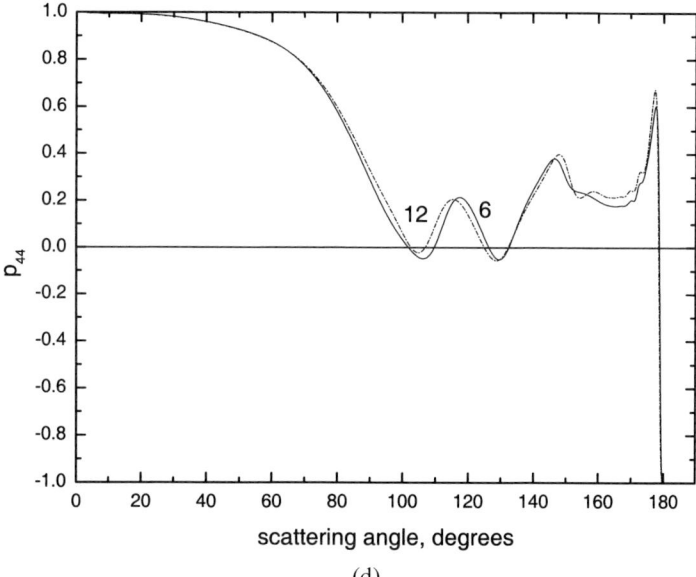

(d)

Figure 3.2 (*cont.*). (*c*) The same as in (*a*) but for the degree of polarization p_l; (*d*) the same as in (*a*) but for p_{44}.

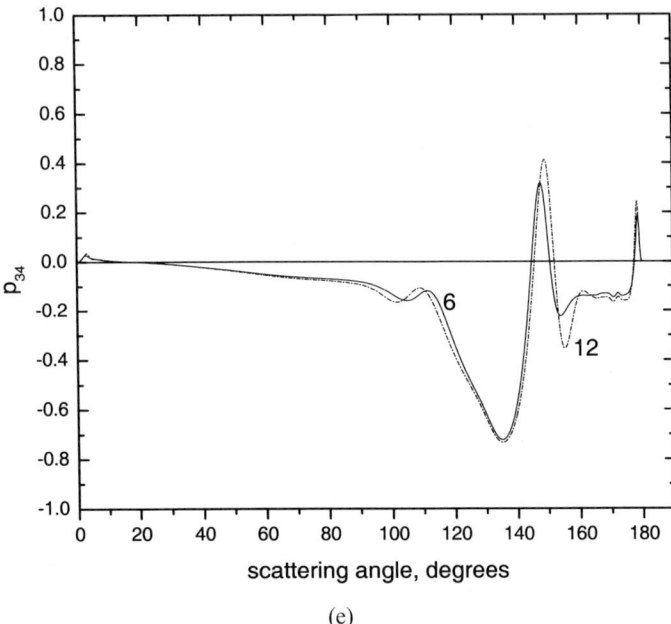

(e)

Figure 3.2 (*cont.*). (*e*) The same as in (*a*) but for p_{34}.

As was specified above, single scattering and absorption characteristics of spherical polydispersions are determined by the refractive index $m = n - i\chi$ of particles and the particle size distribution $f(a)$. Therefore, it is interesting to understand the influence of $f(a)$ and m on the phase function, normalized matrix elements, and parameters q and s. We consider this problem now, using Mie calculations for spherical polydispersions, characterized by the gamma distribution (1.5). Note that it was found that spherical polydispersions, having different PSDs but close values of the effective radius a_{ef} and effective variance Δ_{ef}, have also close values of light scattering characteristics (Hansen and Travis, 1974). Therefore, our study is quite general.

We select the case $a_{ef} = 6\,\mu m$, $\mu = 6$ (see Equation 1.5), which was proposed by Deirmendjian (1969) for the water clouds microphysics characterization as the basic one. Then we change a_{ef} and μ to see the change in scattering characteristics at $\lambda = 0.55\,\mu m$. It is assumed that $m = 1.333$ in Figures 3.1 and 3.2. In Figures 3.3 and 3.4 we keep a_{ef} and μ constant, and study the influence of the change in real and imaginary parts of the refractive index on light scattering characteristics. Let us now analyse the data given in Figures 3.1–3.4.

The phase function, normalized matrix elements, and parameters q and s change drastically with the size of the particles. In particular, the case of $a_{ef} = 0.06$, which is given in Figure 3.1(a) is very close to the Rayleigh scattering, as discussed above. With growth of the particles the phase function becomes more and more extended in

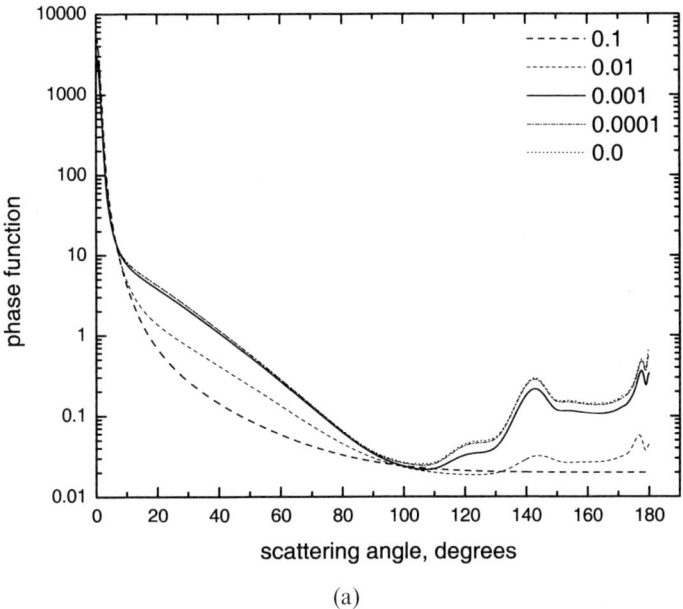

(a)

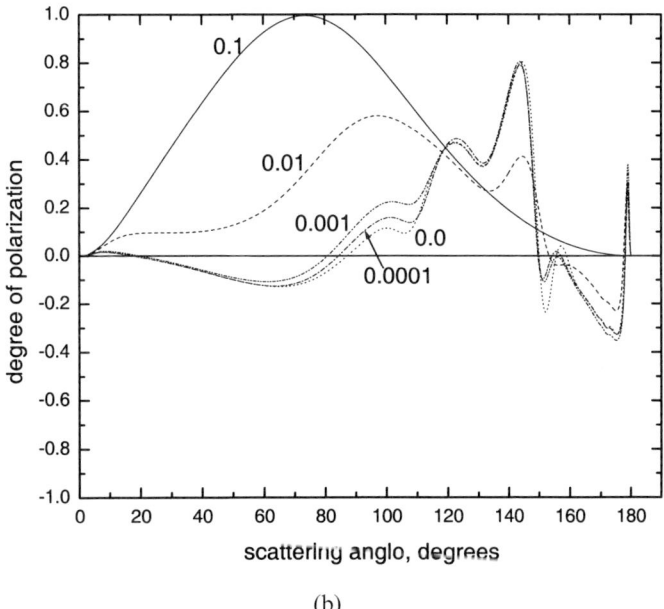

(b)

Figure 3.3. (a) The dependence of the phase function of spherical polydispersions with the PSD (1.5) at $a_{ef} = 6\,\mu m$ and $\mu = 6$ on the imaginary part of the refractive index $\chi = 0$, 0.001, 0.01, and 0.1. It is assumed that $\lambda = 0.55\,\mu m$ and $m = 1.333$; (b) the same as in (a) but for the degree of polarization p_l.

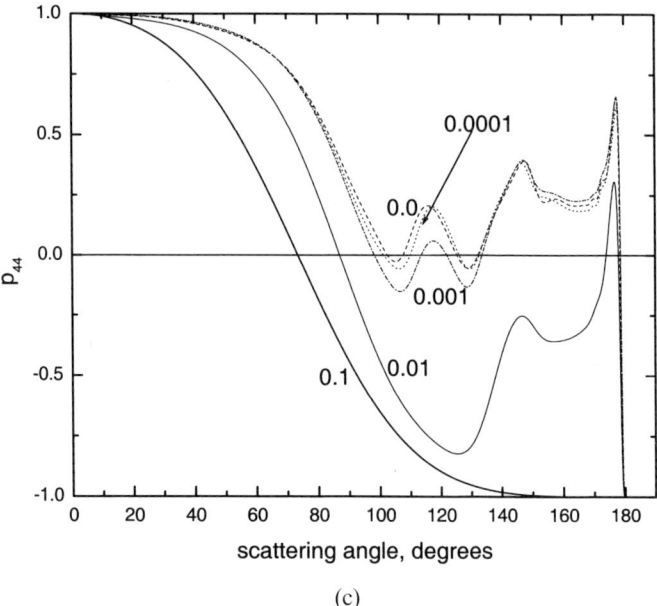

(c)

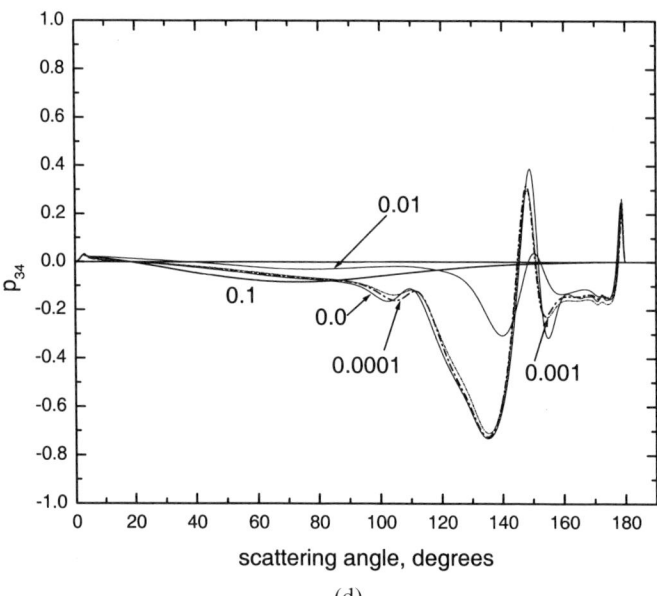

(d)

Figure 3.3 (*cont.*). (c) The same as in (a) but for p_{44}; (d) the same as in (a) but for p_{34}.

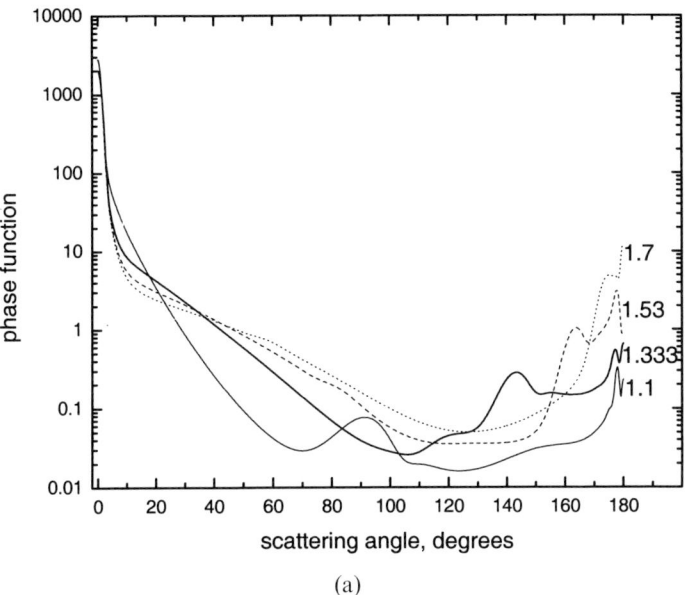

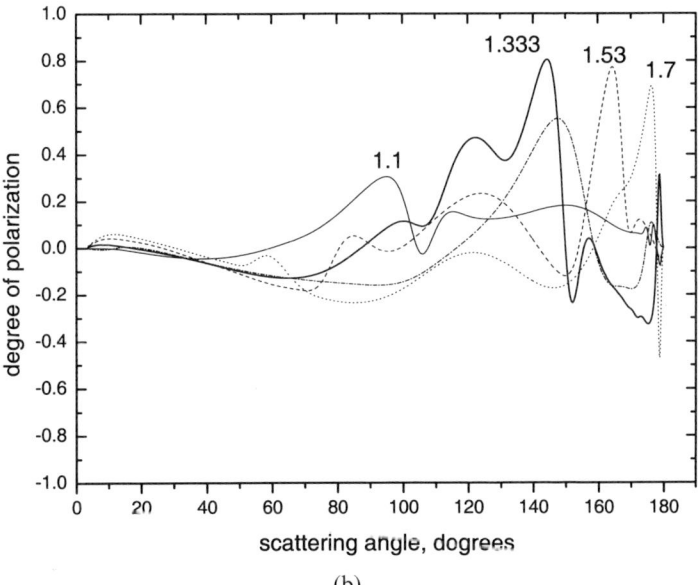

Figure 3.4. (a) The dependence of the phase function of spherical polydispersions with the PSD (1.5) at $a_{ef} = 6\,\mu\text{m}$ and $\mu = 6$ on the real part of the refractive index n equal to 1.1, 1.333, 1.53, and 1.7. It is assumed that $\lambda = 0.55\,\mu\text{m}$ and $\chi = 0$; (b) the same as in (a) but for the degree of polarization p_l.

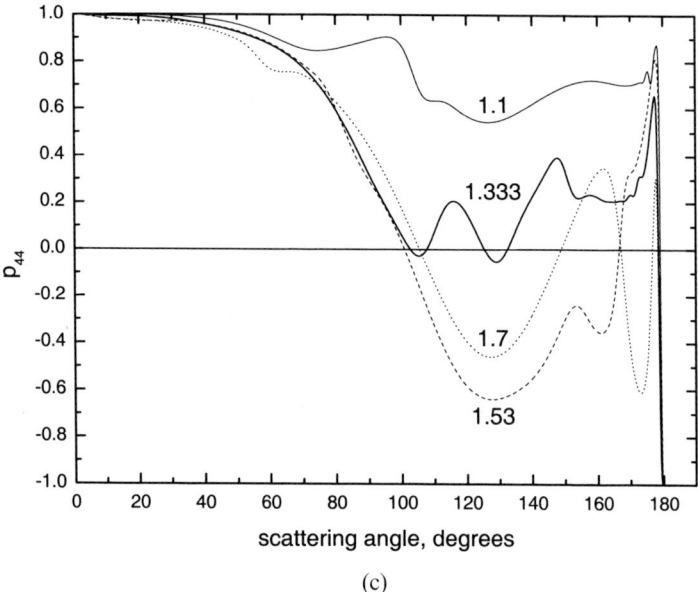

(c)

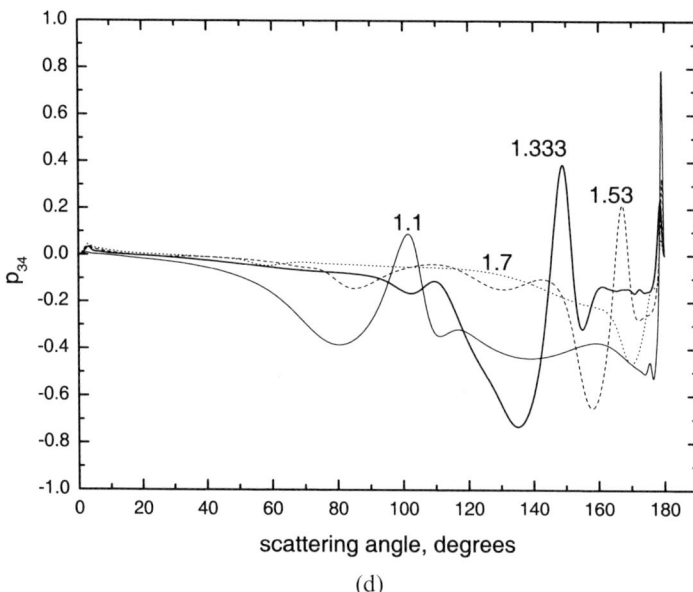

(d)

Figure 3.4 (*cont.*). (c) The same as in (a) but for p_{44}; (d) the same as in (a) but for p_{34}.

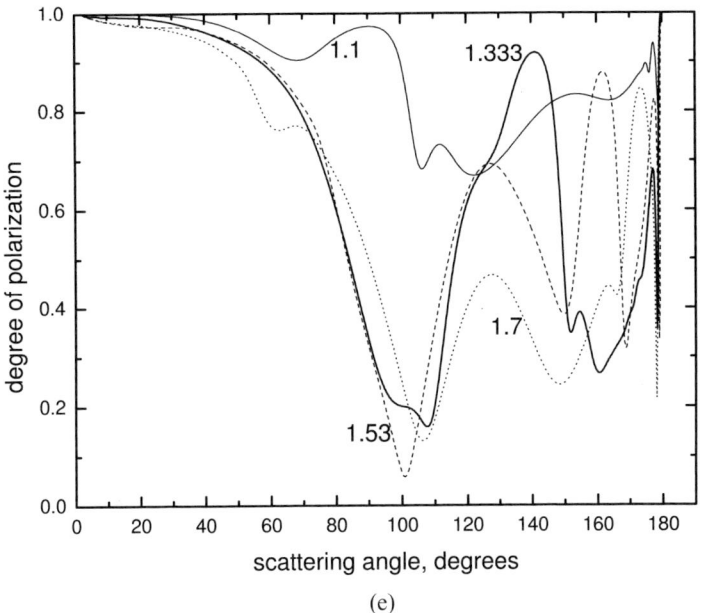

(e)

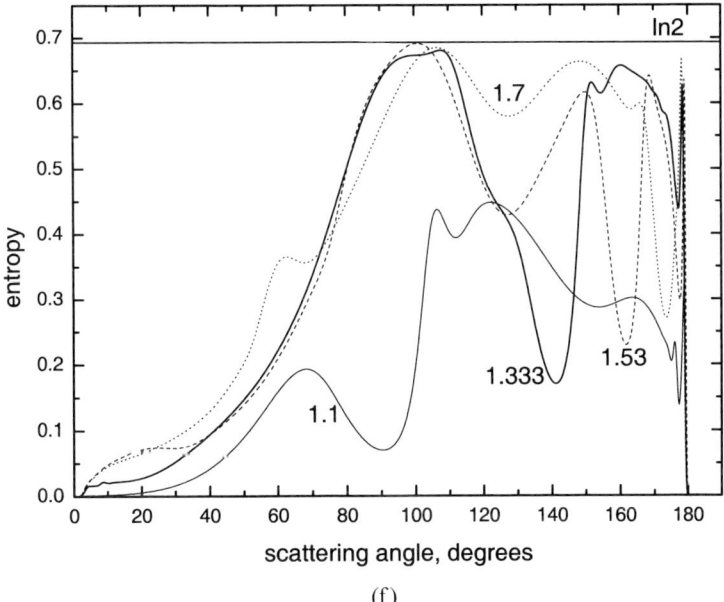

(f)

Figure 3.4 (*cont.*). (e) The same as in (a) but for q; (f) the same as in (a) but for s.

the forward scattering direction. The behaviour of the phase function at small angles is very sensitive to the size of the particles. This feature is used for particle size distribution retrieval on a routine basis. Particles having the effective radius $a_{ef} = 0.6\,\mu\text{m}$, which is close to the wavelength, have phase functions, differing considerably from the Rayleigh scattering phase function. In particular, note the maxima in Figure 3.1(a) around scattering angles 160° and 180°, which are predecessors of the rainbow (near 140°) and glory (near backward direction). These scattering features are clearly visible for particles with effective radii of 6 and 15 μm.

The oscillations of light intensity due to the glory scattering move to larger angles with the size of droplets (or the inverse wavelength). It produces coloured rings around scattering angle 180° for solar illumination of a cloud. This is often observed from an airplane. The radius of the rings can be used for the estimation of a_{ef}.

To the left from the main rainbow located near 140°, we can see the secondary rainbow ($\theta \approx 125°$) at $a_{ef} = 15\,\mu\text{m}$. This rainbow is absent for the case 0.6 μm and only slightly visible at $a_{ef} = 6\,\mu\text{m}$ (see Figure 3.1(a)). The minimum around 130° is called Alexander's band. This corresponds to a region of relatively low light scattered intensity, located between two maxima. Note that positions of the maxima (both for the primary and secondary rainbows) differ at a_{ef} being equal to 6 and 15 micrometers. This can be used for optical particle sizing.

It is known that the phase function depends mostly on the effective size parameter $x_{ef} = ka_{ef}$, where $k = 2\pi/\lambda$ is the wavenumber (at least under the assumption that the refractive index is constant in the spectral band studied). Because the maximum of the primary rainbow moves out of the backward direction to smaller angles with the size of particles (x_{ef} increases), one can expect that the same happens with the inverse wavelength (x_{ef} increases). This will produce a well known dispersion of light by water droplets, illuminated by a white light. Red rings (larger wavelengths), therefore, should be positioned at larger scattering angles. This also follows from Figure 3.1(a).

We emphasize that geometrical optics gives the following values for the rainbow positions: $\theta \approx 137.9°$ for the first rainbow and $\theta \approx 129.1°$ for the secondary rainbow at $m = 1.333$. The discrepancy of these results with data obtained from the wave optics in Figure 3.1(a) indicates that geometrical optics fails at the rainbow for $x_{ef} \leq 170$. This is in contrast to scattering in the vicinity of 40°, where geometrical optics gives a correct value of the phase function even at $x_{ef} \sim 10$ (see Figure 3.1(a)).

Actually, geometrical optics gives infinite values for the light intensity at the rainbow angle. The Airy theory of the rainbow can explain the angular distribution of light in the vicinity of the rainbow angle, but it is only valid for x_{ef} larger than approximately 2000. Airy calculations show that the geometrical optical result for the rainbow scattering angle θ_R does not give the exact position of the maximum intensity in the rainbow. For monodispersed spheres, the maximum occurs at (Newton, 1982):

$$\theta = \theta_R + \upsilon x^{-2/3}$$

where

$$v = 1.0188 \left(\frac{3}{2}\right)^{2/3} \frac{(4-n^2)^{1/6}}{\sqrt{n^2-1}}$$

or $v = 1.73$ at $n = 1.333$. The width and the separation of separate rainbow peaks are proportional to $x^{-2/3}$.

Note also the scattered light intensity oscillation to the right from the rainbow maximum both for cases of $a_{ef} = 6$ and $15\,\mu m$. This oscillation moves to the backward direction for smaller particles. This feature is due to the supernumerary bows present in the scattering diagrams of monodispersed droplets illuminated by a monochromatic light. These bows arise from the interference of light waves with different phases emerging from water droplets. A distribution of particles sizes and white light illumination generally suppress supernumerary bows. However, they often remain visible in the degree of polarization (see Figure 3.1(b)). This also can be used for optical particle sizing.

The normalized phase matrix element $-p_{12}$, which coincides with the degree of linear polarization of singly scattered light p_l at the illumination by unpolarized light, is given in Figure 3.1(b). We see that the scattering phenomenon is an important source of light polarization. Small particles with $a_{ef} = 0.06\,\mu m$ produce a bell-shaped curve with the maximum ($p_l \approx 1$) at the scattering angle close to $90°$. This angle should be exactly equal to $90°$ for Rayleigh particles. We see, therefore, that small deviations from Rayleigh scattering move the angle of the maximal polarization to larger scattering angles. Also the value of p_l at maximum diminishes with the size of droplets. This transformation is well-represented by the curve at $a_{ef} = 0.6\,\mu m$. Note also a broad negative (and small) polarization branch in the forward hemisphere at $a_{ef} = 0.6\,\mu m$. The negative polarization means that light is predominantly linearly polarized in the scattering plane. This feature is also present in data for larger particles with $a_{ef} = 6$ and $15\,\mu m$. It is explained by the fact that light scattered in the forward hemisphere is mostly due to rays twice refracted by water droplets. It is known that transmitted light is generally polarized in the incidence plane, which coincides with the scattering plane in the case considered ($p_l < 0$). This follows from the Fresnel equations (see Appendix 5).

The dependencies $p_l(\theta)$ at $a_{ef} = 6$ and $15\,\mu m$ are generally similar to each other with all oscillations being more pronounced for larger particles. Usually cloud effective radii are in the range $6-15\,\mu m$. Therefore, differences between these cases indicate the most frequent range of variability of curves $p_l(\theta)$ in terrestrial clouds. Rainbow, glory, and supernumerary bows are more pronounced in the curve of degree of polarization than in the intensity of scattered light. They are highly dependant on the size of particles (see especially supernumerary bows located around $143°$ at $a_{ef} = 15\,\mu m$ and close to $153°$ at $a_{ef} = 6\,\mu m$).

It follows from Figure 3.1(b) that the primary rainbow maximum is located at $\theta \approx 140°$ for $a_{ef} = 15\,\mu m$. The value of p_l at this maximum is close to 90%. Such a high polarization is explained by the fact that the primary rainbow is due to one internal reflection in the droplet (and two refractions), and the internal reflection angle for the rainbow scattering is close to the Brewster angle. The secondary rainbow scattering angle is around 125° at $a_{ef} = 15\,\mu\text{m}$. The positions of the primary and secondary rainbows at $a_{ef} = 6\,\mu\text{m}$ are approximately 142 and 120° respectively. We see that the primary and secondary rainbows move closer with increasing particle size. Their separation $\Delta\theta$ should be equal to approximately 8.8° according to the geometrical optics limit. The dependence $\Delta\theta(a_{ef})$ can be used for optical particle sizing.

Let us now analyse the pronounced maximum of the degree of polarization at the scattering angle 95° at $a_{ef} = 15\,\mu\text{m}$ and the similar maximum around 100° at $a_{ef} = 6\,\mu\text{m}$. The degree of polarization at this angle is lower than in the case of the primary and secondary bow. There is a temptation to classify this maximum as a rainbow of the third order. However, this interpretation is not correct. As a matter of fact, the geometrical optics predicts an infinite number of rainbows. However, their intensities are very low in comparison with the main and secondary rainbows. Also we have for the scattering angle at the third rainbow (Bohren and Huffman, 1983):

$$\theta = \pi - 2\sqrt{\frac{m^2 - 1}{15}} + 6\arccos\left[\sqrt{1 - \frac{1}{m^2} + \frac{m^2 - 1}{15}}\right] \quad (3.75)$$

or approximately 68° at $m = 1.333$.

The maximum near 95° is now explained. According to Snell's law, rays do not penetrate into the droplet at incidence angles larger than 90°. We have for the critical angle ψ:

$$\psi = \arcsin(n^{-1}) \quad (3.76)$$

or $\psi = 48.6°$ at $n = 1.333$. This corresponds to the scattering angle $\theta_c = 82.79°$. Therefore, directly transmitted light (after two refractions in the particle) does not contribute to the scattering diagram (and the degree of polarization) at angles larger than approximately 83°. However, contribution of this directly transmitted light into the total scattered energy amounts to 88.45% (if diffraction is neglected). This explains the small values of the phase function around this angle (see Figure 3.1(a)). The degree of polarization at $\theta > \theta_c$ is mostly due to the reflection of light from the droplet's surface. The distribution of reflected energy as the function of the incidence angle ϕ for various refractive indices of the interface droplet-host medium is given in Appendix 5. We see that the polarization of the reflected light is positive. The angle of the maximum of polarization (100%) is given by the Brewster angle $\phi_B = \arctan(n)$ or approximately 53.12° at $n = 1.333$. Then the scattering angle is equal to 73.76° ($180° - 2\phi_B$). Therefore, if light could not penetrate into the droplet and refract there, the distribution of the degree of polarization would be large and positive. This is indeed the case (see Figure 3.3(b)) for large strongly absorbing droplets. Because a large portion of refracted photons contributes to the scattered light in the forward scattering zone (up to $\theta = \theta_c$), this positive polarization is completely lost. However, the refracted light component (if we do not account for reflections inside a droplet) disappears at $\theta > \theta_c$, and we are mostly left with the reflection phenomena.

Summing up, the maximum around 95° for large droplets is explained by the transfer from the refraction-dominated (negative) to the reflection-dominated (positive) pattern of the degree of polarization (see Appendix 5). The decrease of $p_l(\theta)$ in the range of scattering angles 95–115° for the effective radius 15 μm is due to the corresponding decrease in the value of p_l for an externally reflected component.

The behaviour of $p_l(\theta)$ in the rainbow scattering region is explained by the influence of internally reflected and refracted components, which produce strong intensities (and polarization) in the vicinity of rainbows.

Note also the non-monotonic dependence of the degree of polarization near the scattering angle of 180°. The degree of polarization changes sign from a negative to positive value with θ in the glory scattering region (175–180°). It is zero in the forward and backward directions. Therefore, both unpolarized and polarized incident light do not change their state of polarization due to scattering at these directions. This is similar to the Rayleigh case. For larger particles, however, there are other scattering angles of zero polarization as well. For particles with $a_{ef} = 0.6$ μm, we have two additional angles of zero polarization (approximately at 110 and 170°, see Figure 3.1(b)).

Note that this zero polarization in these particular directions is approximately conserved for optically thick layers. This is because multiple light scattering is the process which leads to the increase of entropy and decrease of initial polarization of an incident wave. For the effective size 6 μm we have 8 angles of zero polarization in singly scattered light (including forward and backward directions). These are located approximately at 0, 15, 88, 150, 157, 160, 178, and 180° (see Figure 3.1(b)). The number of angles is reduced to just 6 at the effective radius 15 μm. This is mostly due to larger values of the maximal degree of polarization and the smaller oscillation amplitude in the supernumerary bow at $a_{ef} = 15$ μm as compared to the case of $a_{ef} = 6$ μm. Supernumerary bows disappear with further growth of particles.

Note that in the geometrical optics limit the number of angles of zero polarization becomes equal to 4 (Shifrin, 1951). They are located at 0, 10, 68, and 180°. The angle around 68° is due to the transformation from the refraction-dominated to the reflection-dominated pattern. The angle around 10° is due to the fact that the degree of polarization for the refracted component ($p_l(\theta) < 0$) is much smaller than that of the reflected component ($p_l(\theta) > 0$) in the near forward scattering lobe ($\theta \to 0$). Therefore, the degree of polarization is mostly due to reflected light there. For larger angles the degree of polarization of refracted light increases, and the total degree of polarization changes its sign, as seen in Figure 3.1(b).

Let us now consider the element p_{44} (see Figure 3.1(c)). This gives the degree of circular polarization of scattered light for the incident left-handed polarized light beam. It follows from Figure 3.1(c) that the polarization of scattered light is preserved in the forward scattering direction. It reverses sign and becomes completely right-handed polarized for all types of spherical polydispersions at the exact backward direction. It is known that for the Rayleigh particles, the degree of circular polarization p_{44} is equal to zero at 90°. Light is completely linearly polarized in this case. Actually, the dependence $p_{44}(\theta)$ at $a_{ef} = 0.06$ μm is quite close to this case. However, the zero value of p_{44} occurs at a slightly larger scattering angle.

Particles with $a_{ef} = 0.6\,\mu m$ produce a quite different dependence of $p_{44}(\theta)$ as compared to the case of Rayleigh scattering. In particular, the handedness of the polarization is preserved for almost all scattering angles in contrast to the case of Rayleigh scattering, where forward and backward hemispheres have symmetrical patterns with the same polarization but different handedness (see, e.g., the curve at $a_{ef} = 0.06\,\mu m$ in Figure 3.1(c)).

The polarization state changes its handedness in the vicinity of backward scattering (170–180°). Such abrupt changes in the degree of circular polarization at $\theta \approx 180°$ are common for large particles (see Figure 3.1(c)). The cases of $a_{ef} = 0.06\,\mu m$ and $a_{ef} = 0.6\,\mu m$ have only one point of zero polarization (see Figure 3.1(c)). This is in contrast with the case of large particles. There are five zero polarization points (or neutral points) at $a_{ef} = 6\,\mu m$. Seven neutral points are present at $a_{ef} = 15\,\mu m$. We see that the circular polarization studies have special advantages for the optical particle sizing problems.

The element p_{44} shows how the degree of circular polarization of incident light decreases from its value of one in the exact forward direction due to the scattering process. In contrast, the element p_{34} (see Figure 3.1(d)) describes the angular distribution of circular polarization due to illumination of a scattering volume by horizontally polarized light. Therefore, it gives a linear to circular polarized mode conversion due to the scattering process. For Rayleigh particles, p_{34} vanishes. Therefore, such particles cannot transform linearly polarized light in the circularly polarized mode (and vice versa). This is also the case for large non-absorbing particles in the geometrical optics limit. In this respect even particles with $a_{ef} = 15\,\mu m$ are far away from a geometrical optics limit at the wavelength $0.55\,\mu m$. They do not have very low values of p_{34} at most of the scattering angles (see Figure 3.1(d)). The degree of circular right-handed polarization is larger than 70% at the scattering angle 135° and $a_{ef} = 15\,\mu m$ for an incident horizontally linearly polarized light beam (see Figure 3.1(d)). The curve $p_{34}(\theta)$ at $a_{ef} = 6\,\mu m$ is similar in shape to the case of $a_{ef} = 15\,\mu m$. However, the corresponding minima are broader at $a_{ef} = 6\,\mu m$. They are shifted with respect to minima at $a_{ef} = 15\,\mu m$. It is interesting that particles of all sizes approach a zero value of p_{34} at the scattering angle 180° from the region of positive values of the degree of polarization. We see that the element p_{34} is very sensitive to the size of particles. This can be used in optical particle sizing problems.

Let us assume now that the scattering medium is illuminated by completely linearly polarized light with the azimuth angle of 45°. Then the total degree of polarization of scattered light is given by the function $q(\theta)$, shown in Figure 3.1(e).

We clearly see the glory, primary and secondary rainbow, supernumerary bow, and refraction–reflection transformation peak at $a_{ef} = 15\,\mu m$. These features have already been discussed above, so we will not consider them in further detail. However, note that the degree of polarization q for spherical polydispersions could be quite low: it is only 10% at scattering angle of 160° for $a_{ef} = 15\,\mu m$. Remember that it is assumed that incident light is completely polarized in this case. Thus, spherical polydispersions have quite high depolarizing properties at some scattering geometries. This is in contrast with monodispersed spheres, where we always have $q = 1$. The same holds for Rayleigh particles. Note that $q \approx 1$ for the case $a_{ef} = 0.06\,\mu\mathrm{m}$ as well. It follows that angular light scattering by spherical particles can be used to create light with a desired value of $q \in [0, 1]$. Gorchakov (1972) proposed the use of the function $q(\theta)$ of an atmospheric air for identification of weather type.

The entropy s of a scattered beam, which corresponds to the same experimental situation as in Figure 3.1(e) is given in Figure 3.1(f). By definition, the entropy of a completely polarized light beam is equal to zero. Therefore, we have, $s(\pi) = s(0) = 0$. Small particles (see, e.g., the case $a_{ef} = 0.06\,\mu\mathrm{m}$) produce almost no entropy. This is also the case for single scattering by monodispersed spheres. Large particles, however, are very effective in respect to entropy production (especially in the backward scattering zone outside the rainbow and glory regions). Note that multiple light scattering tends to smooth wave optics scattering effects (see, e.g., the peak around 140°), therefore increasing entropy towards its maximum. The same is true for irregularly shaped particles, which do not have rainbow and glory features.

The influence of the half-width of the particle size distribution is not so pronounced as the influence of the effective size of particles (e.g., compare Figure 3.1(a) with Figures 3.2(a) and (b)). This has a drawback because it is difficult to retrieve the half-width parameter from scattering light measurements in the region of parameters $\mu = 2 - 12$. These values of μ correspond to the coefficient of variance Δ of the particle size distribution in the range of approximately 30 to 60%, which is a typical range of variability of Δ in natural clouds. This special feature of cloudy media allows, however, for the simplification of the cloud microphysics representation in radiative transfer problems. Also this gives a reason why cloud retrieval algorithms (Kokhanovsky, 2001) are devoted to the derivation of the effective radius of droplets only (and not the value of Δ).

Note that there is a strong dependence of the degree of polarization p_l on the half-width of the particle size distribution in the region of the supernumerary bows (around 150° in Figure 3.2(c)). This angle is typical for satellite observation (e.g., the sun at 60° and the nadir observation). Therefore, this phenomenon of the sensitivity of supernumerary bows to Δ can be used to retrieve both a_{ef} and Δ from degree of polarization measurements. Some results in this direction have already been obtained (Breon and Goloub, 1998). The degree of circular polarization for the circularly polarized light illumination (p_{44}) is weakly sensitive to the parameter μ of the particle size distribution (see Figure 3.2(d)). Advances in cloud droplet distribution determination can be achieved measuring the normalized matrix element p_{34} (see Figure 3.2(e)), which is sensitive to μ.

Let us consider now the influence of the imaginary part χ of the refractive index. This is given in Figure 3.3. In particular, it follows from Figure 3.3(a) that the change of the refractive index from zero to 0.0001 does not influence the phase function of cloudy media very much. At $\chi = 0.001$ these changes are already clearly visible. In particular, the probability of light scattering in the backward hemisphere decreases. Note that the influence of χ in the forward hemisphere is not so pronounced. Also we see that both rainbow and glory are still present at

$\chi = 0.01$. The increase of χ up to 0.01 leads to a further decrease of scattered light intensity both in the forward and backward hemispheres. The glory and the rainbow become weaker.

Finally, at $\chi = 0.1$, almost all light which penetrates the sphere is absorbed. Therefore, the scattering pattern is determined by the external reflection, which can be easily found from the Fresnel formulae (see Appendix 5). Note that all phase functions almost coincide at $\theta \approx 100°$ (see Figure 3.3(a)). This indicates that the contribution of light refraction by particles at this angle is negligible (even for non-absorbing particles).

The more pronounced influence of χ on the phase function in the backward hemisphere as compared to scattering in the forward lobe is due to larger light paths inside absorbing particles for multiple internally reflected beams as compared to the directly transmitted component (after just two refractions without reflections).

Let us now consider the degree of linear polarization given in Figure 3.3(b). We see that the degree of polarization is more sensitive to small deviations of absorption from zero (see, e.g., the region around 100°). However, dependencies $p_l(\theta)$ are similar at $\chi \leq 0.001$. For larger χ, dependencies $p_l(\theta)$ differ depending on the actual value of χ. They coincide with the polarization of externally reflected light at $\chi = 0.1$.

The decrease of light beam intensity due to light beam propagation through a length equal to the effective diameter $d_{ef} = 2a_{ef}$ is described by the factor $\Upsilon = \exp(-\alpha d_{ef})$, where $\alpha = 4\pi\chi/\lambda$. We have: $\alpha d_{ef} \approx 27$ at $\chi = 0.1$ and $\Upsilon \ll 1$. Therefore, the contribution of light refraction by a droplet can be neglected in this case (see Figure 3.3(b)).

The influence of absorption on the elements p_{34} and p_{44} is given in Figures 3.3(c) and (d). This is similar to trends just discussed for the dependence $p_l(\theta)$. Note that the zero value of p_{44} at $\theta \approx 73°$ approximately coincides with the maximum p_l at $\chi = 0.1$.

Finally, let us consider the dependence of the light scattering characteristics on the real part of the refractive index n (see Figure 3.4). It follows from Figure 3.4(a) that phase functions become increasingly featureless with the refractive index. A similar trend exists as $n \to 1$. Geometrical optics gives for the position of the primary θ_p and secondary θ_s rainbows (Bohren and Huffman, 1983):

$$\theta_p = \pi - 4\Theta_1 + 2\theta_1, \qquad (3.77)$$

$$\theta_s = 6\Theta_2 - 2\theta_1 \qquad (3.78)$$

where

$$\Theta_j = \arcsin\left[\frac{\sin\theta_j}{n}\right], \quad j = 1, 2. \qquad (3.79)$$

$$\theta_1 = \arccos\sqrt{\frac{n^2-1}{3}}, \quad \theta_2 = \arccos\sqrt{\frac{n^2-1}{8}}. \qquad (3.80)$$

Table 3.4. The dependence of $\theta_p, \theta_s,$ and $\Delta\theta$ on n.

n	θ_p, degrees	θ_s, degrees	$\Delta\theta$, degrees
1.1	84.33	41.31	43.02
1.2	113.62	171.76	−58.14
1.312	134.767	134.772	−0.005
1.333	137.92	129.11	8.813
1.53	159.76	87.97	71.79
1.6	165.07	77.06	88.01
1.7	171.07	63.78	107.29
1.8	175.49	52.65	122.84
1.9	178.51	43.24	135.28

We give values θ_p, θ_s, and their difference $\Delta\theta = \theta_p - \theta_s$ in Table 3.4 and Figure 3.5 for various n. Note that the value of n depends on the temperature of droplets. Therefore, functions $\theta_p(n)$ and $\theta_s(n)$ can be used for the measurements of temperature of particles.

It follows from Figure 3.4(b) that the degree of polarization decreases with a decrease of the refractive index. Also the primary rainbow moves to the backward direction with n and disappears at $n \geq 2$. The value of p_{44} (see Figure 3.4(c)) is generally larger for $n = 1.53$ and $n = 1.7$ as compared to the case $n = 1.33$ (and vice versa for p_{34} (see Figure 3.4(d)). It follows from Figure 3.4(e) that the degree of polarization q is a complicated function of the scattering angle for all n. The minimum polarization occurs at $\theta \approx 100°$ and $n = 1.53$. The entropy production (see Figure 3.4(f)) is lower for optically soft particles with n close to one and increases with n. Note that s is very close to $s_{max} \approx \ln 2$ at $\theta \approx 100°$ and $n = 1.53$. Light scattering at this angle for the case studied can be used for the conversion of polarized light in the unpolarized mode.

It is known that the elements of the phase matrix of spherical particles approach their geometrical optics asymptotic pattern as $\lambda/a_{ef} \to 0$. These are given in Figures 3.6 and 3.7 at $n = 1.333$ and $n = 1.52$. For comparison we also present results of the Mie theory, obtained with the PSD (1.5) at the effective radius equal to $15\,\mu m$, $\lambda = 0.55\,\mu m$, and $\mu = 6$. We see that the geometrical optics approximation generally has better accuracy in the forward scattering hemisphere. It improves with n.

The accuracy is lower in the backward hemisphere. This is mostly due to effects of rainbow and glory. These phenomena should be treated by taking account of the wave nature of light.

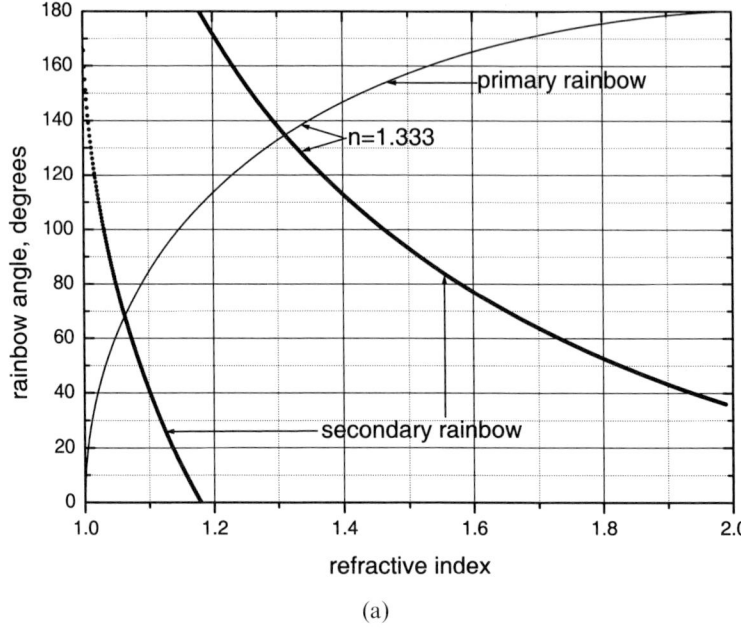

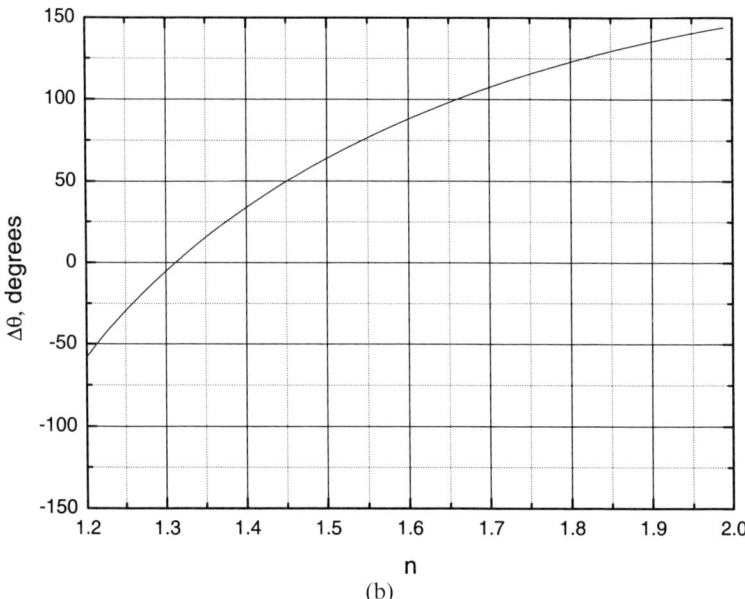

Figure 3.5. (a) The primary and secondary rainbow angles as functions of the refractive index n. It follows that the primary rainbow angle is equal to 138° and the secondary rainbow angle is equal to 129° at $n = 1.333$; (b) the dependence of the separation between primary and secondary rainbows $\Delta\theta$ on the refractive index n.

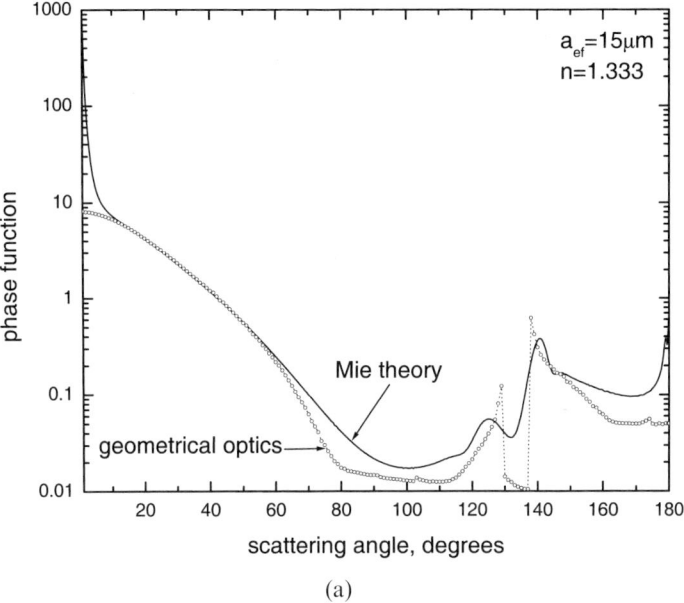

(a)

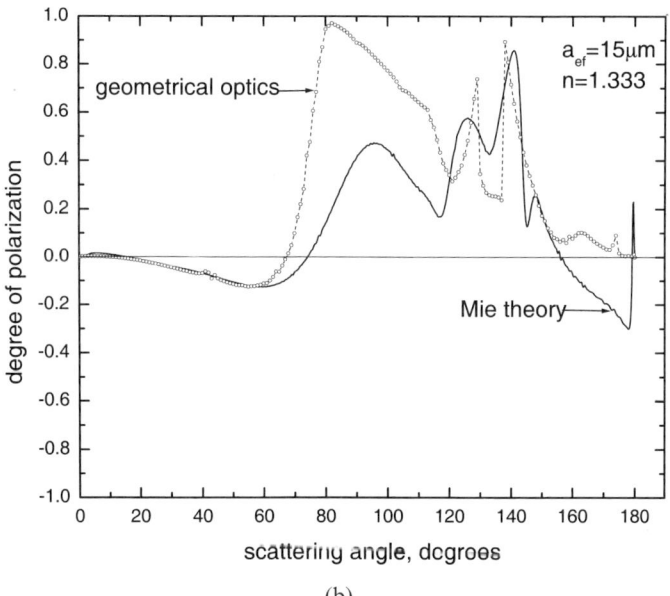

(b)

Figure 3.6. (a) The phase function of spherical polydispersions with the PSD (1.5) at $a_{ef} = 15\,\mu m$, and $\mu = 6$, calculated with the Mie theory and the geometrical optics approach at $\lambda = 0.55\,\mu m$ and $n = 1.333$; (b) the same as in (a) but for the degree of polarization p_l.

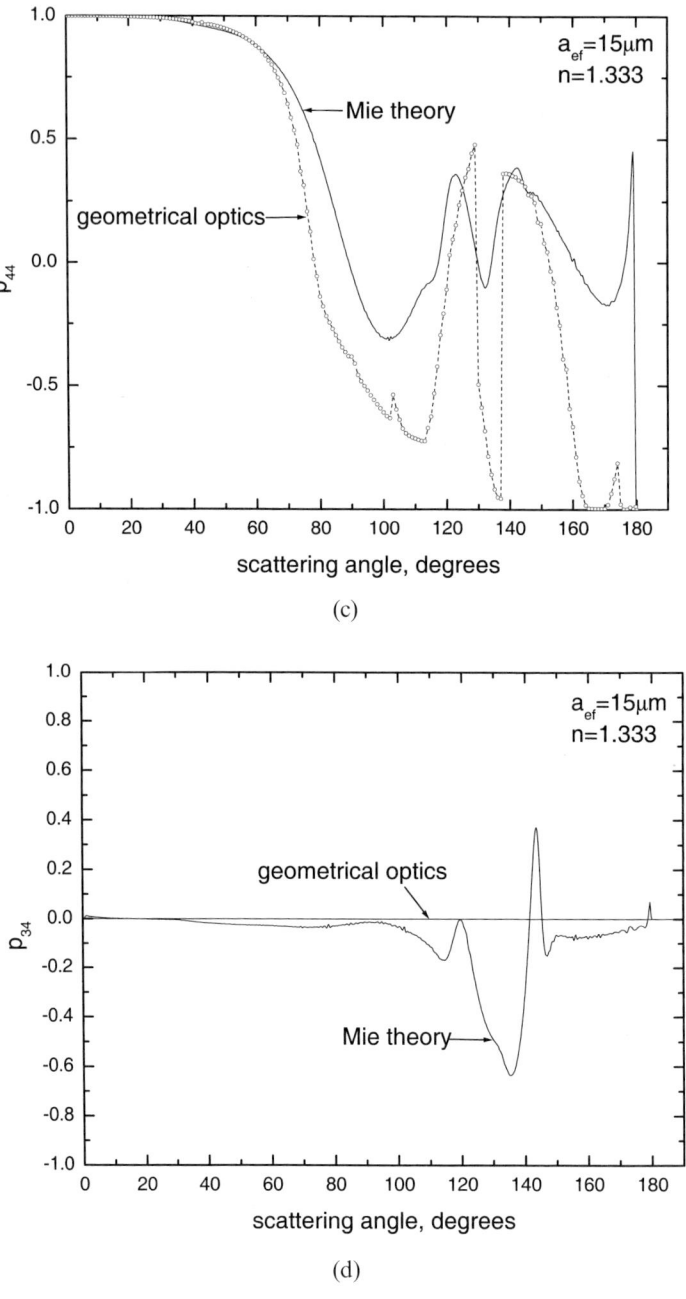

Figure 3.6 (*cont.*). (c) The same as in (a) but for p_{44}; (d) the same as in (a) but for p_{34}.

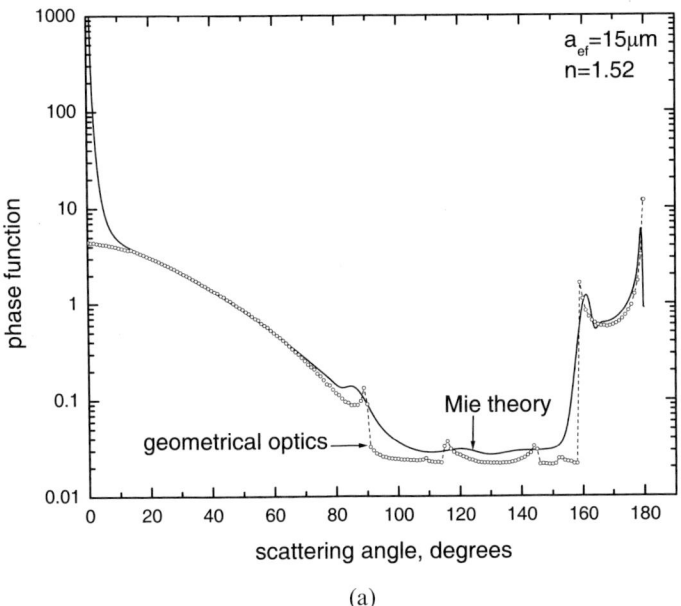

(a)

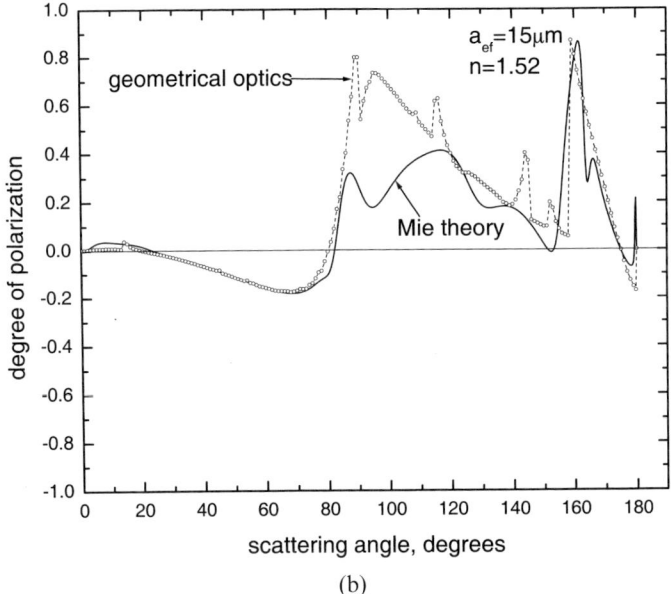

(b)

Figure 3.7. (a) The phase function of spherical polydispersions with the PSD (1.5) at $a_{ef} = 15\,\mu m$ and $\mu = 6$, calculated with the Mie theory and the geometrical optics approach at $\lambda = 0.55\,\mu m$ and $n = 1.52$; (b) the same as in (a) but for the degree of polarization p_l.

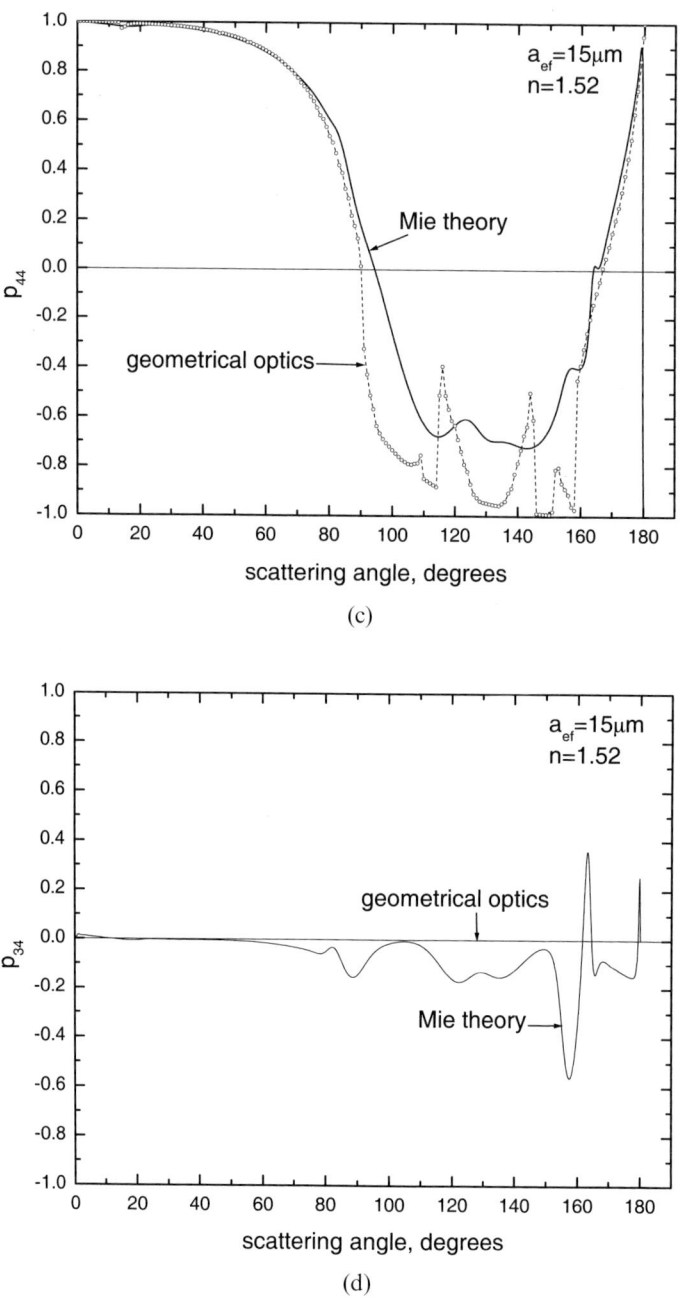

Figure 3.7 (*cont.*). (c) The same as in (a) but for p_{44}; (d) the same as in (a) but for p_{34}.

3.2.3 Bubbles

Let us now consider the peculiarities of light scattering and polarization by spherical particles with the refractive index smaller than that of a host medium. The classical example is an air bubble in water (Davis, 1952). Then, the relative refractive index of a particle n is close to 0.75. For convenience, we will call these particles bubbles even if they are not bubbles in the usual sense of this word (e.g., a water droplet or a bacterium contained in a solid, having a larger refractive index).

The comparison of the phase function and phase matrix elements for air bubbles in water and water droplets in air with the same particle size distribution is given in Figure 3.8. Bubbles are usually much larger than the wavelength of the visible light. Therefore, calculations are performed for the effective radius of particles a_{ef} equal to 15 µm. Note that a_{ef} is defined as a ratio of the third to the second moment of the particle size distribution $f(a)$. It is assumed that the wavelength λ is equal to 0.55 µm. The refractive index of water and air was taken to be 4/3 and 1, respectively. The spherical polydispersions are characterized by the particle size distribution (1.5) at $\mu = 6$.

Results of calculations of the phase function $p(\theta)$, which gives a conditional probability of light scattering in the direction specified by the scattering angle θ are given in Figure 3.8(a). We see that $p(\theta)$ is different for bubbles and droplets not only quantitatively, but also qualitatively. In particular there is no rainbow scattering for bubbles as opposed to droplets (see the angle range 120–160°). There is a small increase in the phase function as $\theta \to \pi$ for bubbles, but this is not so pronounced for droplets. The probability of photon scattering by bubbles in the region of angles from 120 to 180° is considerably smaller compared to droplets. So the backscattered signal from bubbles is comparatively weak. On the other hand bubbles are stronger scatterers at lateral angles (from 40 to 110°). Due to droplets and bubbles being the same size in our calculations, their phase functions coincide as $\theta < \theta_d$ where $\theta_d \approx 5°$ in the case studied. The region of angles $[0, \theta_d]$ specifies the diffraction zone.

There is an interesting feature in the behaviour of the phase function of bubbles at the scattering angle close to 70°. This is related to the fact that rays incident on spherical particles (with relative refractive indices $n < 1$) at angles larger than the critical angle $\varphi = \arcsin(n)$ do not penetrate inside particles. They are totally externally reflected. The critical scattering angle θ_c is related to the critical angle φ: $\theta_c = \pi - 2\varphi = 2\arccos(n)$. This gives for the critical scattering angle: $\theta_c \approx 82.82°$ at $n = 0.75$. Light scattering at $\theta_d < \theta < \theta_c$ is mostly due to the total internal light reflection from the surfaces of bubbles.

The next quantity we would like to consider is the degree of total polarization q, which is exactly equal to one for monodispersed spheres. The difference $\Delta = 1 - q$, therefore, shows the influence of the polydispersity on the value of the degree of total polarization q. It follows from Figure 3.8(e) that the value of q remains larger than 0.9 for bubbles for almost all scattering angles (except the glory scattering region, where it drops to the value of 0.4). It is essentially smaller for droplets with three distinctive minima located at approximately 85, 118, and 160°. For the last

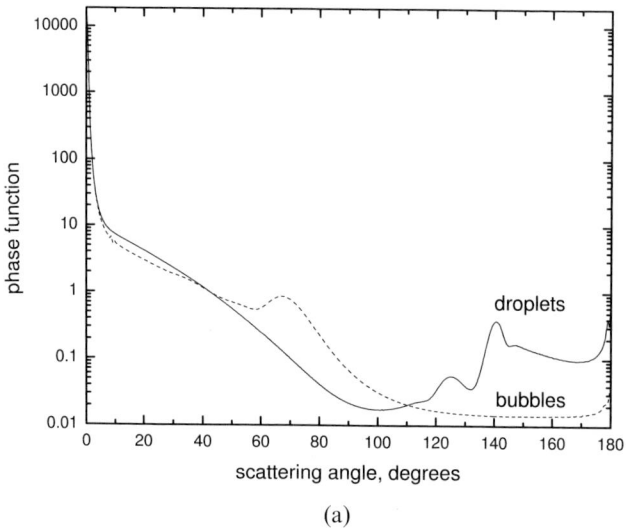

(a)

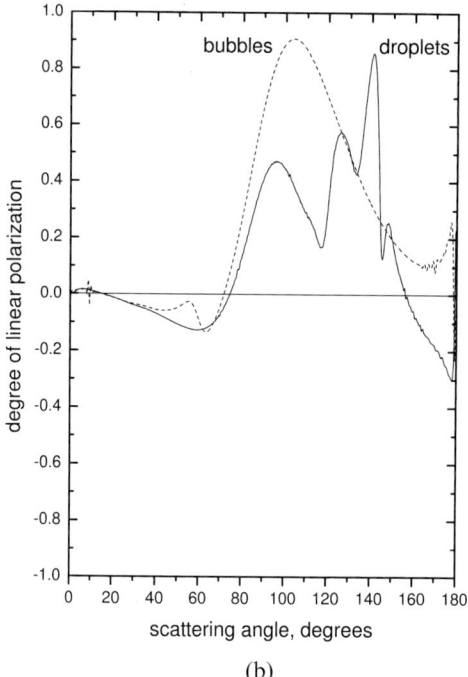

(b)

Figure 3.8. The phase function of spherical polydispersions of bubbles and droplets with the PSD (1.5) at $a_{ef} = 15\,\mu m$ and $\mu = 6$, obtained with the Mie theory. It is assumed that the wavelength $\lambda = 0.55\,\mu m$. The case of bubbles corresponds to $n = 3/4$. For droplets it is assumed that $n = 4/3$; (b) the same as in (a) but for the degree of polarization p_l.

Sec. 3.2] Spherical particles 77

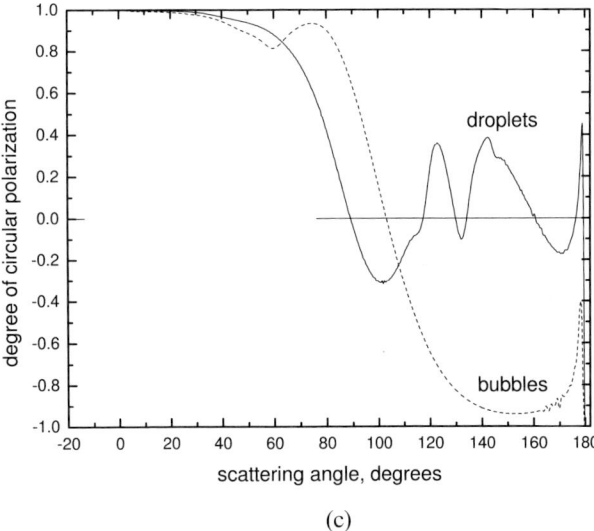

(c)

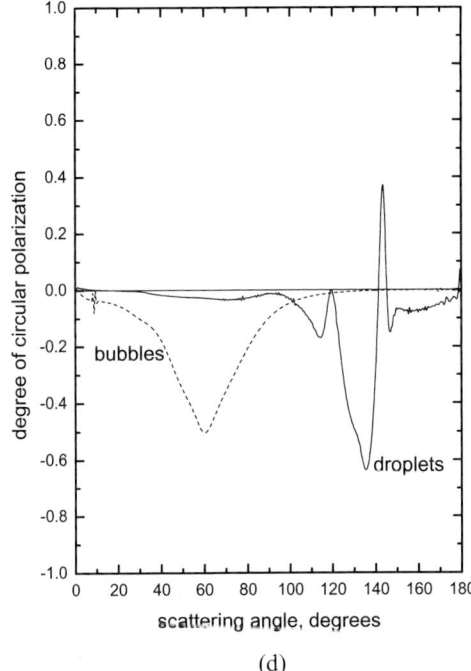

(d)

Figure 3.8 (*cont.*). (c) The same as in (a) but for p_{44}; (d) the same as in (a) but for p_{34}.

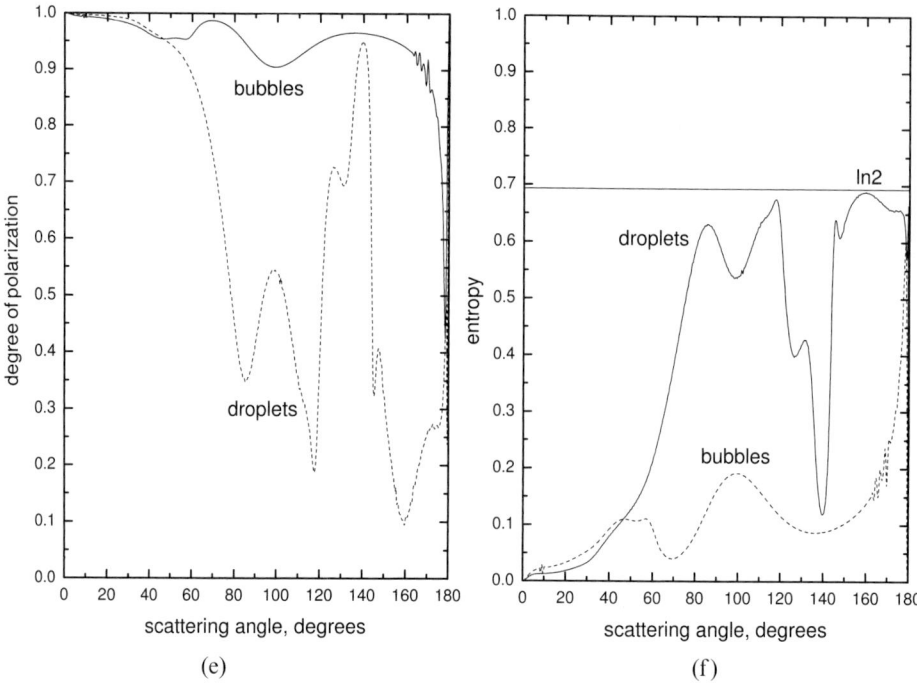

Figure 3.8 (*cont.*). (e) the same as in (a) but for q; (f) the same as in (a) but for s.

minimum, the value of q drops below 0.1. Therefore, at this angle around 90% of incident completely polarized light transforms into unpolarized mode.

The value of q determines the entropy s of the scattered light beam. The dependence $s(\theta)$ is given in Figure 3.8(f). We see that the entropy production generally increases with the scattering angle. It drops at the rainbow (for droplets) and glory regions due to high polarizing abilities of particles there. The entropy production is equal to zero at exact forward and backward scattering regions. Spherical particles do not depolarize incident completely polarized light there.

The degree of linear polarization of initially unpolarized incident light is equal (with opposite sign) to the element p_{12}: $p_e = -p_{12}$. This element is given in Figure 3.8(b). We see that the degree of linear polarization p_l is rather small both for bubbles and droplets in the forward scattering region (below 90°). It increases for larger angles and is larger for bubbles everywhere in the backward hemisphere, except regions where rainbow (around 140°) and glory (around 180°) scattering occurs for water droplets. Interestingly, the value of p_l for bubbles is positive in the backward hemisphere (except at a narrow angle range around 180°). It is negative, however, for droplets at scattering angles larger than 155°. For droplets it becomes positive again around the backward ($\theta \approx \pi$) scattering direction. We see that both bubbles and droplets show the change of the degree of polarization sign at the glory scattering region. Generally the angular dependence $p_l(\theta)$ is much smoother for bubbles as compared to droplets.

We present elements p_{44} and p_{34} in Figures 3.8(c) and (d) respectively. They are equal to the degrees of circular polarization of singly scattered light for different illumination conditions. The case given in Figure 3.8(c) (p_{44}) corresponds to the incident completely right-hand circularly polarized light ($p_c = 1$). We see, therefore, that both bubbles and droplets almost do not change the degree of polarization p_c for angles smaller than 20°. They both convert incident right-hand ($p_c = 1$) circularly polarized light into left-hand ($p_c = -1$) circularly polarized light at the scattering angle of 180°. The curve $p_c(\theta)$ is much more smooth for bubbles compared to droplets. It crosses the line $p_c = 0$ only one time (at $\theta = 105°$) in contrast to the case of droplets, where the degree of circular polarization of scattered light p_c vanishes at seven scattering angles in the backward hemisphere, therefore, producing partially linearly polarized light ($p_c = 0$).

Bubbles change the sense of rotation of the electric vector at angles larger than 105° ($p_c < 0$). In contrast, droplets are characterized by different signs of $p_c(\theta)$, depending on the specific scattering angle interval in the backward hemisphere (see Figure 3.8(c)).

The degree of circular polarization $p_c^*(\theta)$ of scattered light for incident linearly polarized light (the azimuth angle $\psi = -45°$) coincides with the matrix element p_{34}, which is given in Figure 3.8(d). This value is close to zero for bubbles in the backward hemisphere. It is in the range $[-0.65, 0.4]$ for droplets in the same angular range (see Figure 3.8(d)).

The value $p_c^*(\theta)$ for droplets is close to zero in the forward scattering hemisphere compared to quite large values of p_c^* for bubbles (see Figure 3.8(d)). As it follows from Figure 3.8(d), droplets give large negative values of p_c^* at the rainbow angle.

The values of the asymmetry parameters g both for bubbles and droplets for the case given in Figure 3.8(a) are rather close (≈ 0.85). This means that thick layers of both water clouds and air bubbles in water of the same optical thickness will have close values of the diffuse transmittance and reflection coefficients (Kokhanovsky, 2001). This is in contrast to differences in singly scattered light characteristics presented in Figure 3.8(a).

We present results of geometrical optics calculations for the phase function of bubbles (valid as $(a/\lambda) \to \infty$) as compared to the case of the wave optics calculations of $p(\theta)$ at $a_{ef} = 15\,\mu m$ in Figure 3.9(a). This comparison suggests that the cases $a_{ef} = 15\,\mu m$ and $a_{ef} = 30\,\mu m$ are very close to the asymptotic geometrical optics scattering regime in the backward hemisphere. However, there are differences in the region of angles around the scattering angle θ_c, where the total external reflection takes place. The differences as $\theta \to 0$ in Figure 3.9(a) are due to the different sizes of particles assumed for the geometrical optics ($a_{ef} = 1000\,\mu m$) and Mie theory ($a_{ef} = 15\,\mu m$) calculations. The size and the wavelength dependent feature in light scattering by bubbles at angles close to 80° give colour images, if sufficiently small bubbles are illuminated by white light. This is similar in appearance to the rainbow phenomenon for droplets, but it takes place in the forward hemisphere and has a different physical origin. Correspondent caustics are also different.

Let us now consider how the light scattering patterns of bubbles change with refractive index (see Figure 3.10). To simplify, we will assume that there is no

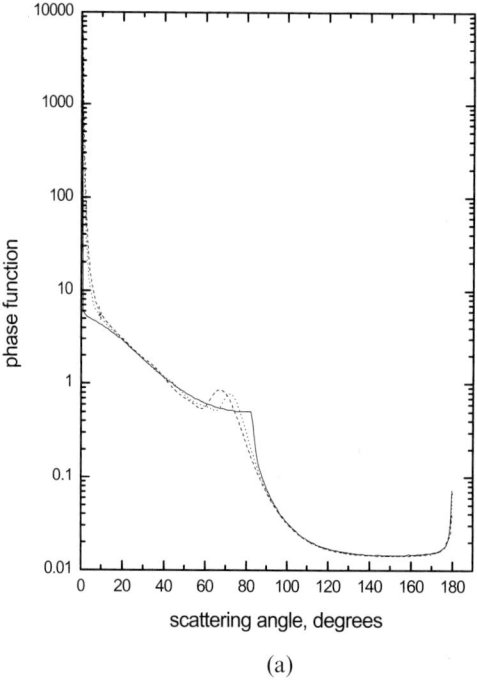

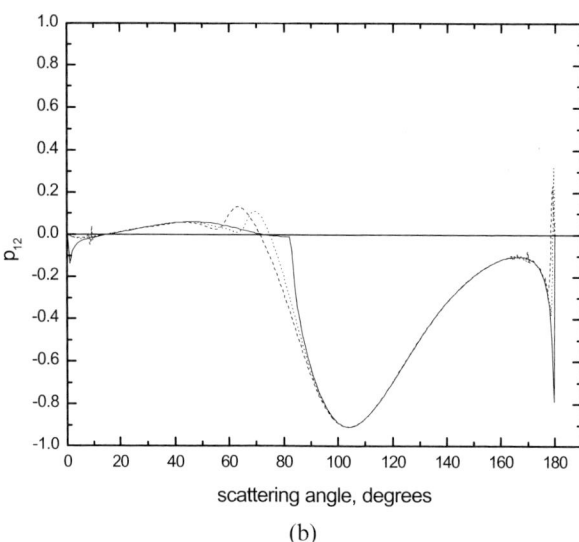

Figure 3.9. (a) The phase function of spherical polydispersions of bubbles and droplets with the PSD (1.5) at $a_{ef} = 15\,\mu\text{m}$ (long-dashed curve) and $a_{ef} = 30\,\mu\text{m}$ (short-dashed curve), obtained with the Mie theory. It is assumed that the wavelength $\lambda = 0.55\,\mu\text{m}$, $\mu = 6$, and $n = 3/4$. The results of the geometrical optics calculations are given by a solid line; (b) the same as in (a) but for p_{12}.

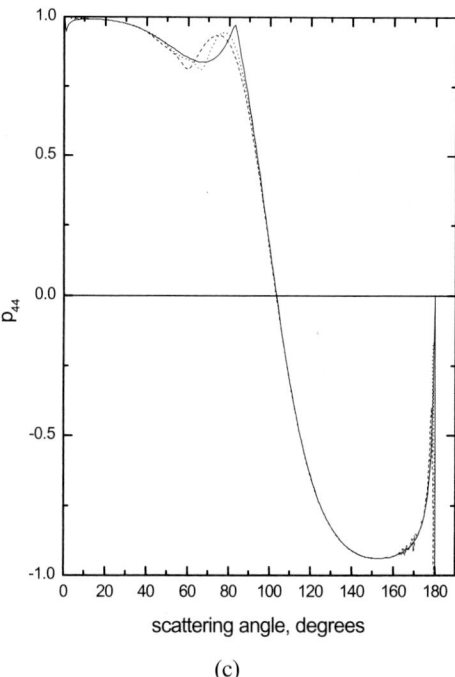

(c)

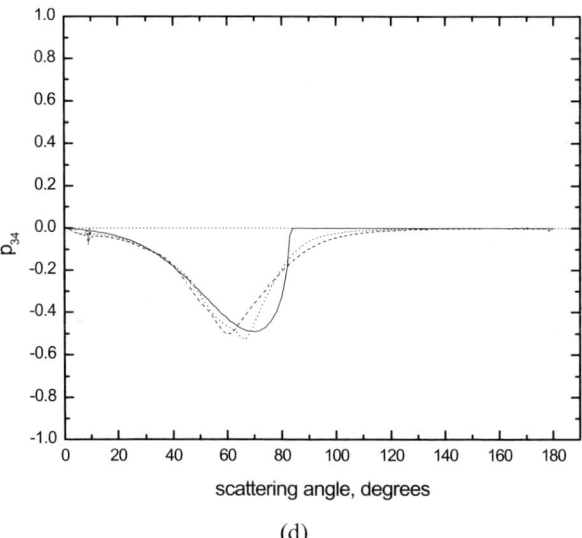

(d)

Figure 3.9 (*cont.*). (c) The same as in (a) but for p_{44}; (d) the same as in (a) but for p_{34}.

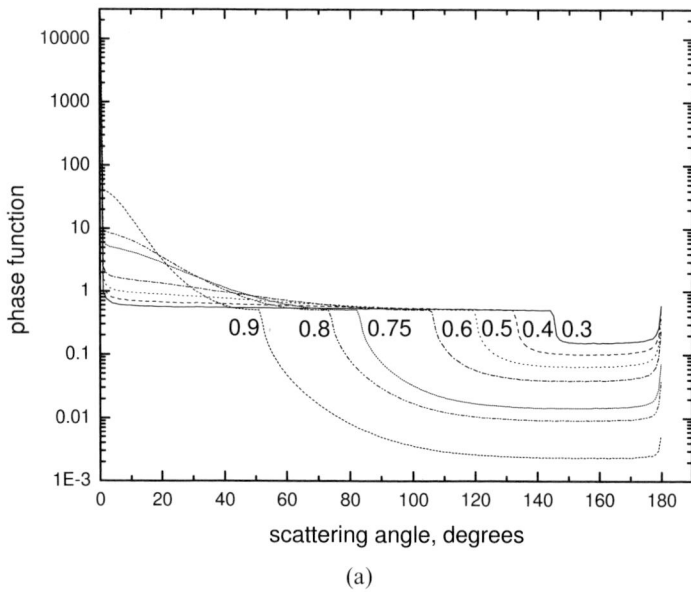

(a)

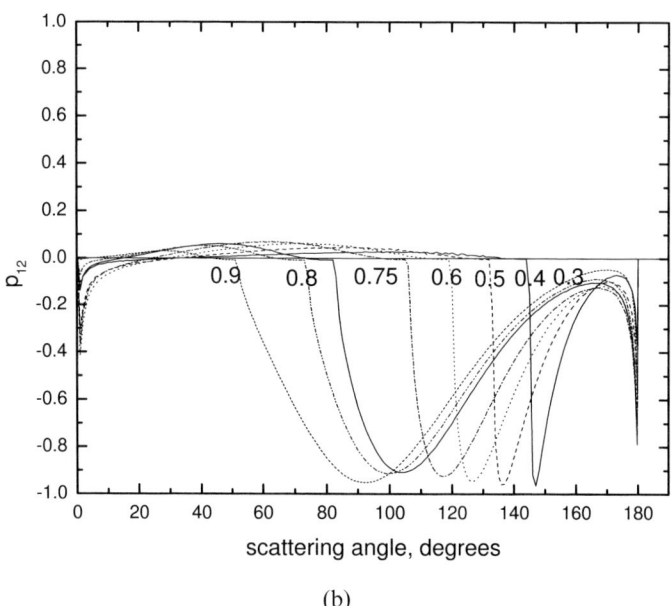

(b)

Figure 3.10. (a) The phase function of spherical non-absorbing bubbles obtained in the framework of the geometrical optics approximation at the refractive index equal to 0.3, 0.4, 0.5, 0.6, 0.75 (air bubbles in water), 0.8, and 0.9. It is assumed that the wavelength $\lambda = 0.55\,\mu m$ and the radius of bubbles is equal to 1 mm; (b) the same as in (a) but for p_{12}.

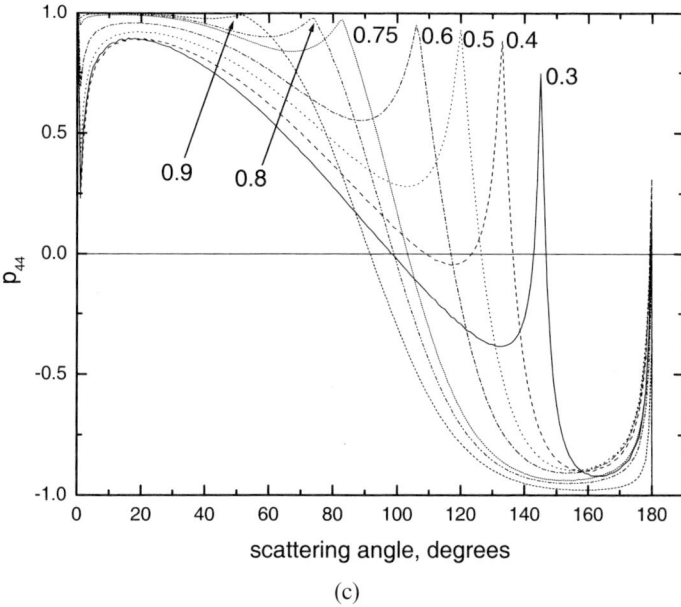

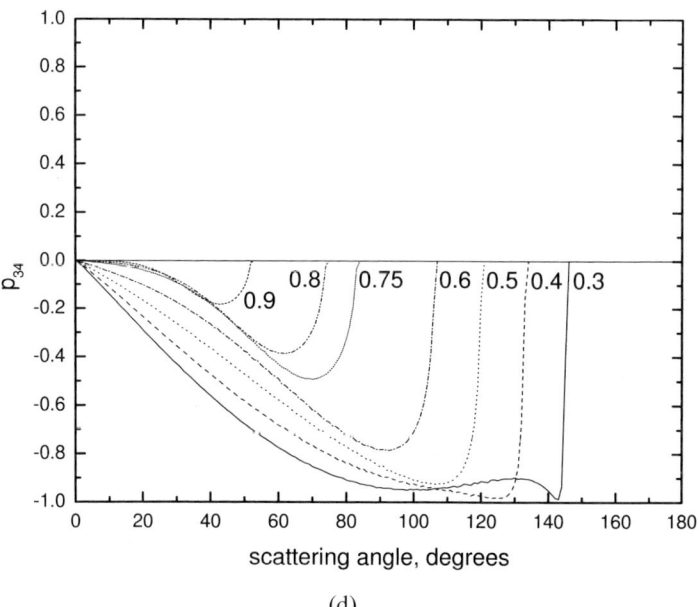

Figure 3.10 (*cont.*). (c) The same as in (a) but for p_{44}; (d) the same as in (a) but for p_{34}.

Table 3.5. The asymmetry parameter g for various n smaller than 1.

n	g
0.0	0.5
0.3	0.5380
0.4	0.5807
0.6	0.7160
0.75	0.8443
0.8	0.8869
0.9	0.9614
1.0	1.0

absorption of light by the particles or the host medium. Also, we assume that the bubbles are much larger than the wavelength of the incident light. Then normalized scattering characteristics are size-independent and are given by geometrical optics equations.

The phase function $p(\theta)$, calculated using the geometrical optics approach, is given in Figure 3.10(a) for various relative indices n of monodispersed bubbles ($n = 0.75$ corresponds to the case of air bubbles in water). We assumed in the calculations that the radius of the bubble was $a = 1$ mm and $\lambda = 0.55\,\mu$m. So we have $\xi \equiv (a/\lambda) \approx 1818$, which explains a sharp peak in the forward direction (its width is proportional to λ/a). As was stated above, $p(\theta)$ does not change with ξ (outside of the diffraction peak) in the framework of the geometrical optics approach for nonabsorbing particles. Thus, the results given in Figure 3.10(a) correspond to the asymptotic scattering regime ($\xi \to \infty$).

The behaviour of phase functions is completely different for angles $\theta < \theta_c$ and $\theta > \theta_c$, where $\theta_c = 2\arccos(n)$. The value of θ_c grows with n^{-1}. In particular, we see that for $n=0.3$ the phase function is almost constant in the region of scattering angles smaller than $145°$ (with the exception of the diffraction zone at small angles). It sharply decreases at $\theta = \theta_c$. Then it is again almost constant and increases as $\theta \to 180°$ (the axial focusing effect). The sharpness of the step at $\theta = \theta_c$ increases with $\Delta n = 1 - n$.

The forward–backward asymmetry of the phase functions is usually represented by the asymmetry parameter g. This is given in Table 3.5 (see also Figure 3.11) for various values of n.

We see that g increases as $n \to 1$ as it should be. The solid curve in Figure 3.11 gives a fit, using the Gaussian distribution:

$$g = g_0 + \frac{A}{\Delta\sqrt{\pi/2}} \exp\left\{-2\left(\frac{n - n_0}{\Delta}\right)^2\right\}, \qquad (3.81)$$

where $g_0 = 0.49274$, $n_0 = 1.04882$, $\Delta = 0.69233$, and $A = 0.44466$. Note that the

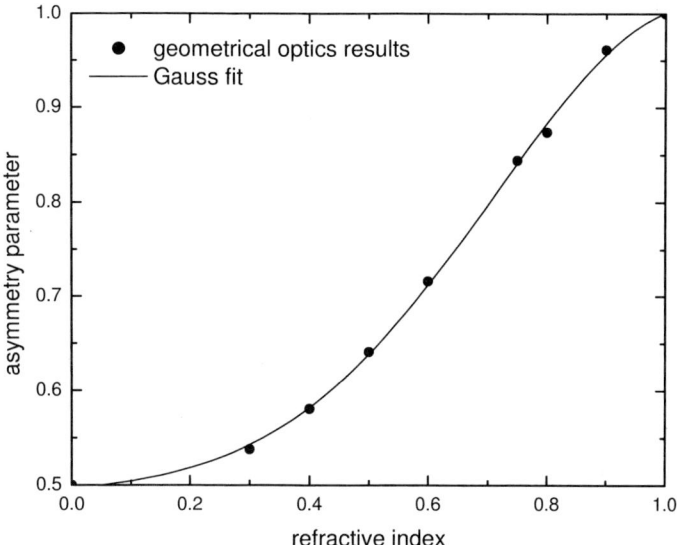

Figure 3.11. The dependence of the asymmetry parameter g on the refractive index of bubbles n, obtained using geometrical optics approximation. Solid line corresponds to Equation 3.81.

parameter g determines the reflectance $r = 1 - t$ and the transmittance t of optically thick bubbly layers. In particular, we have for the large optical thickness τ of a bubbly medium (Kokhanovsky, 2001):

$$t = \frac{1}{1.07 + 0.75\tau(1-g)}. \quad (3.82)$$

Bubbly and cloudy media with the same values of $\tau(1-g)$ have close values of r and t as $\tau \to \infty$. This is in contrast with quite different single scattering laws (see Figure 3.8(a)).

The elements of the matrix $\hat{p}$ of bubbles are presented in Figures 3.10(b)–(d). The element p_{12}, given in Figure 3.10(b), is equal (with opposite sign) to the degree of linear polarization p_l of scattered light, assuming that the incident light is unpolarized. Therefore, we see that p_l is mostly positive (the oscillations of the electric vector are in the plane perpendicular to the scattering plane, containing incident and scattered beams). The degree of linear polarization sharply increases at the critical scattering region. Then it has a maximum and again decreases for larger scattering angles. The maximum is sharper for smaller n. Note that there is also an increase in the degree of linear polarization in the vicinity of the backscattering direction. Also we have: $p_l(0) = p_l(\pi) = 0$ for all n.

The element p_{44}, presented in Figure 3.10(c), gives the degree of circular polarization p_c of scattered light, assuming that incident light is completely right-hand circularly polarized. In particular, we see that $p_c = -1$ for scattered light at exactly backward direction for arbitrary n. Generally, scattering by bubbles reduces the circular polarization of incident circularly polarized light. In particular, the value of p_c is equal to zero for four scattering angles at $n = 0.3$. The strong variation of p_c is observed near the backward direction. For air bubbles in water ($n = 0.75$) p_c varies very little with $\theta < \theta_c$. However, it drops sharply at the critical angle, reversing the sign of p_c around $\theta \approx 105°$. Therefore, the scattering process at angles larger than $105°$ for air bubbles in water transforms incident right-hand circularly polarized light to a left-hand circularly polarized mode.

The element p_{34}, presented in Figure 3.10(d), gives the degree of circular polarization value of p_c for the case of linearly polarized incident light with the azimuth $-45°$. We see that the value of p_{34} is equal to zero at $\theta > \theta_c$. Therefore, linearly polarized light cannot be transformed in a circularly polarized mode at $\theta > \theta_c$ in the geometrical optics scattering regime.

Following this discussion of various matrix elements, we would like to underline that data given in Figure 3.10 can be used to study the interaction of arbitrarily polarized light beams with bubbly media in the framework of the vector radiative transfer equation. In particular, it could be of importance for oceanic water, where air bubbles are present in large numbers.

3.3 ELLIPSOIDAL PARTICLES

3.3.1 General remarks

The scattering of light by a spherical particle can be studied in great detail, using Mie theory as shown above. The scattering behaviour depends only on the relative refractive index m and the size parameter ka. This is the reason why this theory is most often used for the interpretation of experiments and theoretical studies.

However, the spherical model cannot describe many essential features which appear when dealing with non-spherical particles (e.g., dust aerosols, ice crystals). For instance, for spheres, $p_{22} = 1$ and $p_{44}/p_{33} = 1$ for all scattering angles. This is not supported by experiments performed for collections of non-spherical particles (i.e., Volten et al. (2001) observed values of p_{22} in the range 0.3–1.0, depending on the scattering angle).

Let us now consider the case of ellipsoidal particles. To describe the optical properties of such a particle we need to know (apart from its refractive index) three semi-axes a, b and c. Also the orientation of the particle with respect to the incident beam should be taken into account. This adds three more parameters (the Euler angles φ, ψ, and ζ). Moreover, the scattered light parameters in all scattering directions $\vec{n}(\vartheta, \phi)$ differ. So instead of a single scattering angle θ we have to consider the zenith angle ϑ and the azimuthal angle ϕ. Thus, instead of just three parameters m, ka, and θ we have nine parameters m, ka, kb, kc, φ, ψ, ζ, ϑ, and ϕ (m is a complex number).

Theoretical studies become very much involved in this case. This is not only due to the multi-parametric nature of the problem, but also due to the fact that simple solutions like those given by Mie theory are not available for non-spherical particles. All calculations are time-consuming. Moreover, they cannot be performed at all in the case of large average size parameters of the ellipsoids (e.g., $ka = 300$). In this case the geometrical optics can be applied, but the accuracy of the geometrical optics approach cannot be properly checked. Note that for spherical particles the geometrical optics approach cannot describe rainbow scattering. Similar problems arise for non-spherical particles in the range of $ka = 200$–1000, where geometrical optics is not accurate (e.g., for rainbow scattering) and the wave optics theory encounters computational problems. Note that for some light scattering characteristics and angular ranges geometrical optics already has high accuracy at $ka = 50$–100. This can partially solve the problem. However, the extension of the exact theory to larger sizes (e.g., up to $ka = 1000$) remains unsolved for ellipsoidal particles at present.

The situation is complicated by the fact that non-spherical particles rarely (or never) occur in a single shape mode. Take an ice crystal cloud, for example. There, one finds particles of various forms, with no possibility of seeing two particles which are exactly the same. But at least the refractive index is the same. For dust aerosols and oceanic hydrosols, not only the shape but also the refractive index differ, not to mention the internal structure of particles.

This complexity is responsible for the existence of many different models of light scattering in complex media with non-spherical particles. Some of them are very crude and rely primarily on spherical scattering models. Other more sophisticated models are based on the representation of actual disperse systems, using models of ellipsoidal, cylindrical, or other regular shapes or combinations thereof. Theoretical models of irregularly shaped particles have also been developed.

Perhaps, the most simple and well-studied model of a non-spherical particle (except for the case of an infinitely long cylinder, which has a very narrow range of applicability) is that of an ellipsoid. Many important features of light scattering by non-spherical particles (but certainly not all) can be described by this model (especially taking into account various distributions of ellipsoids by their shapes, sizes, orientations, and refractive indices).

3.3.2 Rayleigh ellipsoids

Let us now consider the case of randomly oriented ellipsoids. In the general case we have 5 parameters (m, ka, kb, kc, and θ) of the problem as compared to 3 parameters (m, ka, θ) for the case of spheres. Interestingly enough, in the particular case of Rayleigh ellipsoidal particles ($|mkc| < |mkb| < |mka| \ll 1$, $kc < kb < ka \ll 1$) parameters ka, kb, and kc combine in one number y as far as the phase matrix calculations are concerned. This means that the number of free parameters is reduced to just three as for the case of spheres. This simplifies the analysis considerably.

For such small particles the electric field within a particle is constant and is in phase with the external field. Therefore, a particle can be replaced by a single oscillating dipole with the polarizability tensor $\hat{\alpha}$ and the simple theory of the dipole scattering can be applied for the determination of light scattering and absorption characteristics. The components of the polarizability tensor depend on the shape and the internal structure of particles.

For randomly oriented ellipsoids (Van de Hulst, 1981; Bohren and Huffman, 1983):

$$C_{abs} = \frac{-4\pi k}{3} \text{Im}(\alpha_1 + \alpha_2 + \alpha_3), \quad C_{sca} = \tfrac{8}{9}\pi k^4 (|\alpha_1|^2 + |\alpha_2|^2 + |\alpha_3|^2),$$

$$\hat{P}(\theta) = \frac{3}{3y+1} \begin{pmatrix} y + \cos^2\theta & \cos^2\theta - 1 & 0 & 0 \\ \cos^2\theta - 1 & 1 + \cos^2\theta & 0 & 0 \\ 0 & 0 & 2\cos\theta & 0 \\ 0 & 0 & 0 & \frac{1+3y}{3+y}\cos\theta \end{pmatrix},$$

where α_1, α_2, and α_3 are components of the diagonilized polarizability tensor of a particle:

$$\hat{\alpha} = \begin{pmatrix} \alpha_1 & 0 & 0 \\ 0 & \alpha_2 & 0 \\ 0 & 0 & \alpha_3 \end{pmatrix},$$

C_{sca} and C_{abs} are average scattering and the absorption cross sections per particle, $\hat{P}(\theta)$ is the phase matrix,

$$y = A/B, \quad A = \frac{6-M}{5}, \quad B = \frac{2+3M}{5}, \quad M = \frac{\text{Re}(\alpha_1^*\alpha_2 + \alpha_1^*\alpha_3 + \alpha_2^*\alpha_3)}{|\alpha_1|^2 + |\alpha_2|^2 + |\alpha_3|^2}.$$

It follows for spheres: $A = B = y = 1$. The Rayleigh phase matrix $\hat{P}(\theta)$ is defined in the coordinate system attached to the particle. Thus, it should be rotated to be used in the vector radiative transfer equation.

The phase matrix depends on the internal structure of particles, their refractive index and shape. However, it does not depend on the size of particles. The Rayleigh phase function for randomly oriented particles can be presented in the following form:

$$p(\theta) = \frac{y + \cos^2\theta}{y + 1/3}.$$

It should be pointed out that the value of M is in the range $[-0.5, 1]$ (Bohren and Huffman, 1983) and, correspondingly, $y \in [1, 13]$. The phase function at different values of y is presented in Figure 3.12(a). One can see that the non-sphericity of particles leads to more isotropic phase functions. However, it does not move the position of the minimum at the scattering angle of 90°. Interestingly, all the phase functions have two common intersection points; where the phase function is equal to one. It takes place at scattering angles which are equal to $\arccos(1/\sqrt{3})$ and $2\pi - \arccos(1/\sqrt{3})$ (or 54.7° and 125.3°), respectively. Values of the phase

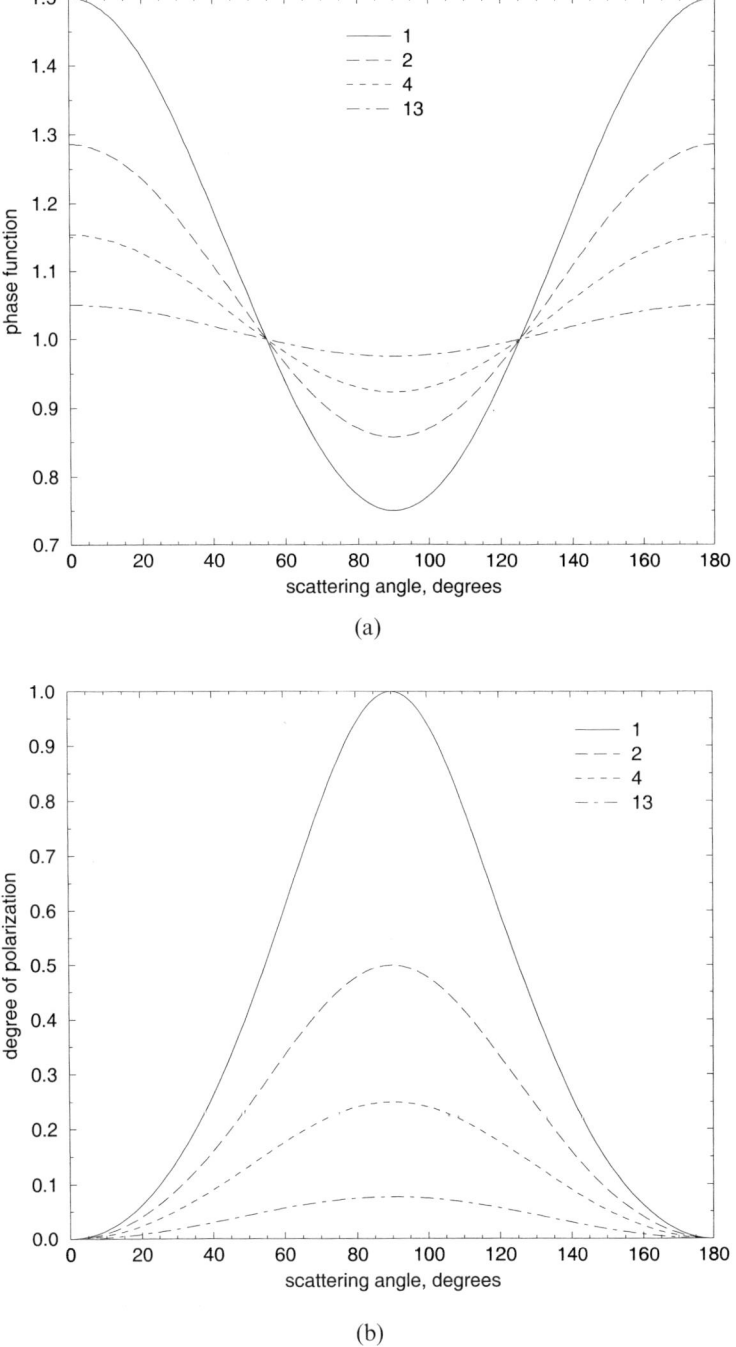

Figure 3.12. (a) The phase function for $y = 1, 2, 4, 13$; (b) the same as in (a) but for the degree of polarization.

functions are identical in the forward and backward scattering directions ($p(0) = p(\pi)$ at any y).

One can see that in the case of Rayleigh scattering the components of the Stokes vector I_s, Q_s, U_s, and V_s of the singly scattered light are proportional to the following functions (the incident light with the intensity I_0 is unpolarized)

$$I_s \sim (A + B\cos^2\theta)I_0, \qquad Q_s \sim B(\cos^2\theta - 1)I_0, \qquad U_s = 0, \qquad V_s = 0.$$

Therefore, the degree of polarization $p_l = -Q_s/I_s$ of the singly scattered light is

$$p_l(\theta) = \frac{1 - \cos^2\theta}{y + \cos^2\theta}.$$

The degree of polarization, obtained with this equation, is presented in Figure 3.12(b). The non-sphericity of the particles reduces the degree of polarization considerably. One can see that the maximum of the degree of polarization at a scattering angle of 90° is inversely proportional to the parameter y: $p_l(90°) = y^{-1}$. For instance, at $y = 2$ then $p(90°) = 0.5$. This feature can be used for the experimental determination of y.

The angle between the direction of polarization and the plane containing the initial and final photon directions (the scattering plane) $\psi = \frac{1}{2}\arctan(U_s/Q_s)$ is equal to 0 or $\pi/2$. The choice of the angle $\psi = 0$ or $\psi = \pi/2$ depends on the sign of the element p_{12} ($\psi = 0$ for $p_{12} > 0$ and $\psi = \pi/2$ for $p_{12} < 0$). For Rayleigh scattering it follows that $p_{12}(\theta) \leq 0$ and $\psi = \pi/2$. Thus, singly scattered light is partially linearly polarized in the plane perpendicular to the scattering plane (the ellipticity $e = 0$).

It is worth pointing out that the asymmetry parameter for Rayleigh particles of any shape is zero and the radiation pressure cross section $C_{pr} = C_{ext} - gC_{sca}$ coincides with the extinction cross section.

To apply the formulae for the values of C_{sca}, C_{abs}, and $\hat{P}(\theta)$ given above to radiative transfer problems one should specify the exact shape and the internal structure of the particles and calculate the values of α_j. Analytical results for the components of the polarizability tensor $\hat{\alpha}$ of uniform ellipsoids and soft ($|m - 1| \ll 1$) Rayleigh particles of any shape are given below. It should be pointed out that the ellipsoidal shape is a very general one. An ellipsoid transforms to a sphere, prolate, or oblate spheroid for a particular choice of axes.

According to the definition, the induced dipole moment of a particle $\vec{p}$ can be described by the following equation (van de Hulst, 1981)

$$\vec{p} = \hat{\alpha}\vec{E}_0,$$

where $\hat{\alpha}$ is the polarizability tensor and $\vec{E}_0$ is the incident electric field. On the other hand, the dipole moment of a particle in an electric field is (van de Hulst, 1957)

$$\vec{p} = \int_V \frac{m^2(\vec{r}) - 1}{4\pi}\vec{E}(\vec{r})\,d\vec{r},$$

where $\vec{E}(\vec{r})$ is the electric field, $m(\vec{r})$ is the refractive index of a particle at the point with the radius vector $\vec{r}$, and V is the volume of the particle. It follows for soft

($|m-1| \ll 1$) Rayleigh particles of any shape to a good approximation: $\vec{E}(\vec{r}) = \vec{E}_0(\vec{r})$ and

$$\alpha_j = \frac{(\langle m^2 \rangle - 1)}{4\pi} V,$$

where

$$\langle m^2 \rangle = \frac{1}{V} \int_V m^2(\vec{r}) \, d\vec{r}.$$

For example, one can obtain for uniform particles with the refractive index m that $\langle m^2 \rangle = m^2$. Thus, for optically soft isotropic uniform Rayleigh particles of any shape it follows:

$$\hat{P}(\theta) = \frac{3}{4} \begin{pmatrix} 1 + \cos^2 \theta & \cos^2 \theta - 1 & 0 & 0 \\ \cos^2 \theta - 1 & 1 + \cos^2 \theta & 0 & 0 \\ 0 & 0 & 2\cos\theta & 0 \\ 0 & 0 & 0 & 2\cos\theta \end{pmatrix},$$

$$C_{abs} = \gamma V, \qquad C_{sca} = \frac{k^4 V^2}{6\pi} |m^2 - 1|^2,$$

where $\gamma = (4\pi\chi/\lambda)$. It is worth remarking that scattering and absorption cross sections of an optically soft Rayleigh particle depend on the volume V of a particle only, irrespective of specific shape. Thus, spherical, cylindrical, and cubic particles of the same volume have the same values of the cross sections C_{abs} and C_{sca}. The elements of the Rayleigh phase matrix $\hat{P}(\theta)$ do not depend on the size, shape, and refractive index of particles in this particular case.

For the specific case of uniform spherical Rayleigh particles at any m (Van de Hulst, 1957)

$$\vec{E}(\vec{r}) = \frac{3}{m^2 + 2} \vec{E}_0$$

and

$$\alpha = \frac{3(m^2 - 1)}{4\pi(m^2 + 2)} V,$$

where it being assumed that the refractive index of the host medium is equal to one. Thus, one can obtain

$$C_{abs} = -3kV \, \mathrm{Im}\left(\frac{m^2 - 1}{m^2 + 2}\right), \qquad C_{sca} = \frac{3}{2\pi} k^4 V^2 \left|\frac{m^2 - 1}{m^2 + 2}\right|^2.$$

The phase matrix as $m \to 1$ is the same as that for soft Rayleigh particles of any shape ($\alpha_1 = \alpha_2 = \alpha_3$ and $M = 1$).

The imaginary part of the refractive index is extremely small in the optical band of the electromagnetic spectrum for most substances in the natural environment ($\chi \ll n$). Thus, it follows approximately in this case that

$$C_{abs} = f(n)\gamma V, \qquad C_{sca} = \frac{3}{2\pi} k^4 V^2 \left(\frac{n^2 - 1}{n^2 + 2}\right)^2,$$

where

$$\gamma = \frac{4\pi\chi}{\lambda}, \qquad f(n) = \frac{9n}{(n^2+2)^2}.$$

One can see that the light absorption by a Rayleigh particle decreases with the refractive index n and vice versa for the value of C_{sca}. The scattering cross section of a particle is proportional to the volume squared and the absorption cross section to the volume.

Another case in which the polarizability tensor of a particle can be calculated by elementary means is that of ellipsoids. The main components α_j of the polarizability tensor of an ellipsoid are (Van de Hulst, 1981):

$$\alpha_j = \frac{V}{4\pi} \frac{m^2-1}{1+L_j(m^2-1)},$$

where for geometrical factors $L_j (L_1 + L_2 + L_3 = 1)$:

$$L_j = \frac{a_1 a_2 a_3}{2} \int_0^\infty \frac{ds}{(a_j^2+s)f(s)}$$

and $j = 1, 2, 3$, $f(s) = \sqrt{(s+a_1^2)(s+a_2^2)(s+a_3^2)}$, $a_1 \geq a_2 \geq a_3$ are the semi-axis of an ellipsoid.

It follows that $\alpha_j \to ((m^2-1)V/4\pi)$ as $m \to 1$ and the polarizability tensor is reduced to the scalar value, which does not depend on the shape of the particles in this case. However, for large and intermediate values of $m-1$, the dependence of α_j on the geometrical factors L_j is very important.

The values of L_j can be represented by elliptic integrals of the first $F(\vartheta, t)$ and the second $E(\vartheta, t)$ kind (Osborn, 1945):

$$L_1 = \frac{\cos\varphi \cos\vartheta}{\sin^3\vartheta \sin^2\beta} \{F(\vartheta,t) - E(\vartheta,t)\},$$

$$L_2 = \frac{\cos\varphi \cos\vartheta}{\sin^3\vartheta \sin^2\beta \cos^2\beta} \left\{ E(\vartheta,t) - \cos^2\beta F(\vartheta,t) - \frac{\sin^2\beta \sin\vartheta \cos\vartheta}{\cos\varphi} \right\},$$

$$L_3 = \frac{\cos\varphi \cos\vartheta}{\sin^3\vartheta \cos^2\beta} \left\{ \frac{\sin\vartheta \cos\varphi}{\cos\vartheta} - E(\vartheta,t) \right\},$$

where

$$\cos\vartheta = \frac{a_3}{a_1}, \quad \cos\varphi = \frac{a_2}{a_1}, \quad \sin\beta = \sqrt{\frac{1-\left(\frac{a_2}{a_1}\right)^2}{1-\left(\frac{a_3}{a_1}\right)^2}} = \frac{\sin\varphi}{\sin\vartheta} = t,$$

$$F(\vartheta,t) = \int_0^\vartheta \frac{d\vartheta}{\sqrt{1-t^2\sin^2\vartheta}}, \qquad E(\vartheta,t) = \int_0^\vartheta \sqrt{1-t^2\sin^2\vartheta}\, d\vartheta.$$

Table 3.6. The values of L_j for different shapes of particles
$$\left(\varsigma = \frac{a_1}{a_3}, \varsigma = \frac{\varsigma + \sqrt{\varsigma^2 - 1}}{\varsigma - \sqrt{\varsigma^2 - 1}}, v = \frac{\sqrt{\varsigma^2 - 1}}{\varsigma}\right).$$

Shape	L_1	L_2	L_3
Prolate spheroid ($a_1 > a_2 = a_3$)	$\dfrac{1}{\varsigma^2 - 1}\left\{\dfrac{\varsigma \ln \varsigma}{2\sqrt{\varsigma^2 - 1}} - 1\right\}$	$\dfrac{1 - L_1}{2}$	$\dfrac{1 - L_1}{2}$
Very slender prolate spheroid ($a_1 \gg a_2 = a_3$)	$\dfrac{1}{\varsigma^2}\{\ln 2\varsigma - 1\}$	$\dfrac{1 - L_1}{2}$	$\dfrac{1 - L_1}{2}$
Needle	0	$\tfrac{1}{2}$	$\tfrac{1}{2}$
Oblate spheroid ($a_1 = a_2 > a_3$)	$\dfrac{1}{2(\varsigma^2 - 1)}\left\{\dfrac{\varsigma^2 \arcsin v}{\sqrt{\varsigma^2 - 1}} - 1\right\}$	L_1	$1 - L_1 - L_2$
Plate	0	0	1
Very flat oblate spheroid ($a_1 = a_2 \gg a_3$)	$\dfrac{\pi}{4\varsigma}\left\{1 - \dfrac{4}{\pi\varsigma}\right\}$	L_1	$1 - L_1 - L_2$
Sphere ($a_1 = a_2 > a_3$)	$\tfrac{1}{3}$	$\tfrac{1}{3}$	$\tfrac{1}{3}$

It follows from these equations that $L_1 + L_2 + L_3 = 1$ as it should be. The general formulas for the values of L_j are simplified for specific shapes (see Table 3.6).

The optical characteristics of small irregular particles are obtained by integrating the corresponding optical characteristics (e.g., scattering and absorption cross sections) over all possible values of L_1 and L_2 weighted by a shape probability function $P(L_1, L_2)$. Such an approach was used by Nevitt and Bohren (1984) for calculation of the backscattering coefficients of irregularly shaped aerosol particles in the infrared (IR) region of the electromagnetic spectrum.

For particles in random orientation one obtains,

$$C_{abs} = \frac{\gamma n V}{3} \sum_{j=1}^{3} \frac{1}{(1 + \nu L_j)^2 + (\mu L_j)^2},$$

$$C_{sca} = \frac{k^4 V^2}{18\pi} \sum_{j=1}^{3} \frac{(\nu + (\nu^2 + \mu^2)L_j)^2 + \mu^2}{[(1 + \nu L_j)^2 + (\mu L_j)^2]^2},$$

where $\gamma = \dfrac{4\pi\chi}{\lambda}$, $\nu = n^2 - 1 - \chi^2$, and $\mu = 2n\chi$.

3.3.3 Spheroidal particles

The case of ellipsoidal particles with arbitrary values of the semi-axes a, b, and c can be treated in the framework of the T-matrix approach or other techniques (Mishchenko et al., 2002). These approaches are restricted to effective size parameters smaller than approximately 200. For larger sizes, geometrical optics ray-tracing methods can be used (Macke and Mishchenko, 1996).

However, even for randomly oriented ellipsoidal particles there is a dependence on two additional parameters (b/a and c/a). This is the reason that in most cases, up to date the models of oblate ($a = b > c$) and prolate ($a > b = c$) spheroidal particles are employed for the interpretation of experimental data. Oblate spheroids are obtained by rotating the ellipse around its small axis c. Prolate spheroids are obtained by rotating it around its large axis a.

Actually, for a spheroid we need to specify only two axes. Let us denote them as $\alpha = |AC|$ and $\beta = |BD|$. Also we assume that the spheroid is obtained by rotation around axis BD. Now we can introduce the axis ratio $\xi = \alpha/\beta$. By definition, this value is larger than 1 for oblate spheroids and it is smaller than 1 for prolate spheroids. We see that in contrast to the case of spherical particles, we have just one additional parameter (the axis ratio ξ), which allows a broad range of shapes to be considered (see Table 3.6), starting from the needle at $\xi \to 0$ (prolate spheroids) to a plate at $\xi \to \infty$ (oblate spheroids). The case of spheres corresponds to $\xi = 1$. Note that the geometrical factors L_j for ellipsoids can be calculated analytically (see Table 3.6).

This model is a further step forward in comparison to a simple model of spherical polydispersions. It can be used to explain many features of light scattering by non-spherical particles (but again – not all!) such as an enhanced scattering at side angles and the cross-polarization effects, etc. Here we come to an important question. Most natural media are composed of non-spherical particles with shapes which are very remote from being spheroids, yet their optical characteristics (e.g., phase functions) can be in principle modelled by spheroids having various two-parametric particle size distribution functions $f(\alpha, \beta)$.

Is there any value in a microphysical model which does not correspond to the real microphysics of a medium? The answer to this question depends on the problem studied. For instance, if one is interested in the detailed shape of particles in a selected volume, then it is better to use well-developed imaging techniques, microscopy, etc. Clearly, light scattering particle characterization techniques have their limitations, and we come to the very end of their applicability if we need to study complex systems with non-spherical particles of various shapes.

Let us take a problem of satellite aerosol remote sensing now (specifically, global aerosol optical thickness retrieval procedures). Here it is of importance to have information on the phase function (and not specifically on the shape of particles) to derive the aerosol optical thickness. This function should represent experimental measurements of the atmospheric phase function. The question of the detailed microstructure of the medium is then outside of the scope of the problem.

To illustrate this, we consider the aerosol optical thickness determination in the framework of the single scattering approximation ($\tau \leq 0.01$), assuming a black underlying surface. Then the reflection function is given by Equation 2.63, i.e.,

$$\mathcal{R} = \frac{\omega_0 p(\theta)}{4(\mu_0 + \mu)} \left\{ 1 - \exp\left[-\left[\frac{1}{\mu} + \frac{1}{\mu_0}\right]\tau \right] \right\},$$

where polarization effects are neglected. It follows from this formula that the aerosol optical thickness τ is determined by the ratio $\mathfrak{R}/p(\theta)$, where $\mathfrak{R}$ is the measured reflection function. The value of ω_0 is close to unity for most cases (Silva et al., 2002). Errors in $p(\theta)$ (and not the exact shapes of particles themselves) will influence the result of the aerosol optical thickness retrieval. Here the notion of the 'optically equivalent medium' appears. This is a highly idealized medium, the local optical properties of which (e.g., phase functions, extinction, and absorption coefficients) coincide with the optical characteristics of the complex medium under study. It is not be possible to introduce an exact equivalency, of course.

However, an approximate equivalency can be introduced and used for the solution of practical problems. This approach has its shortcomings and uncertainties, but it allows a lot of important practical problems to be solved, which could not otherwise.

The results of exact calculations of the phase matrices of spheroidal particles are given in Figure 3.13 for various sizes of particles and axis ratios. The following interesting results can be derived from this figure:

1. The element $p_{22} \to 1$ as $\xi \to 0$, $\xi \to 1$, and $\xi \to \infty$. It differs from unity more considerably in the backward hemisphere than it does in the forward hemisphere. The variation of p_{22} with the scattering angle is more important for larger particles. We have: $p_{22} \to 1$ as $ka \to 0$.
2. Elements of the normalized phase matrix are almost independent of the size of particles as $\xi \to 0$. They only weakly depend on the size as $\xi \to \infty$ (with exception of p_{34}). They tend to Rayleigh scattering diagrams for these asymptotic cases.
3. The angular dependence of elements p_{33} and p_{44} is similar. Values of p_{33} and p_{44} are larger than their values for Rayleigh particles (solid lines are to the left of all other curves) with exception of the case of large aspect ratios. The variability of these elements with size is largest for spherical particles.
4. The degree of polarization $p_l = -p_{12}$ generally increases with the axis ratio ξ and decreases with the size of particles. The maximum of the polarization curve moves into the backward hemisphere with the size of particles.
5. The absolute value of the element $p_{34} \to 0$ as $\xi \to 0$ and $\xi \to \infty$ (with the exception of p_{34} in the forward hemisphere). The largest deviations of p_{34} from zero are for spherical particles. Remember that the element p_{34} indicates the effectivity of the linear polarized light transformation into a circularly polarized mode. It appears that this process is more efficient for spheres.

Note that the conclusions presented here cannot be extrapolated to the case of larger particles without special care. The normalized matrix elements and the phase function for large ($ka \to \infty$) randomly oriented spheroidal particles at $n = 1.5$ and $\xi = 0.05, 0.5, 1, 2$ and 20 are presented in Figure 3.14. The wavelength is equal to 0.55 µm and the semi-axis a is equal to 1 mm for oblate spheroids and 20 mm for prolate spheroids. Because of the large size of the particles, computations were performed using the geometrical optics technique. This explains the sharp features

96 Local optical characteristics [Ch. 3

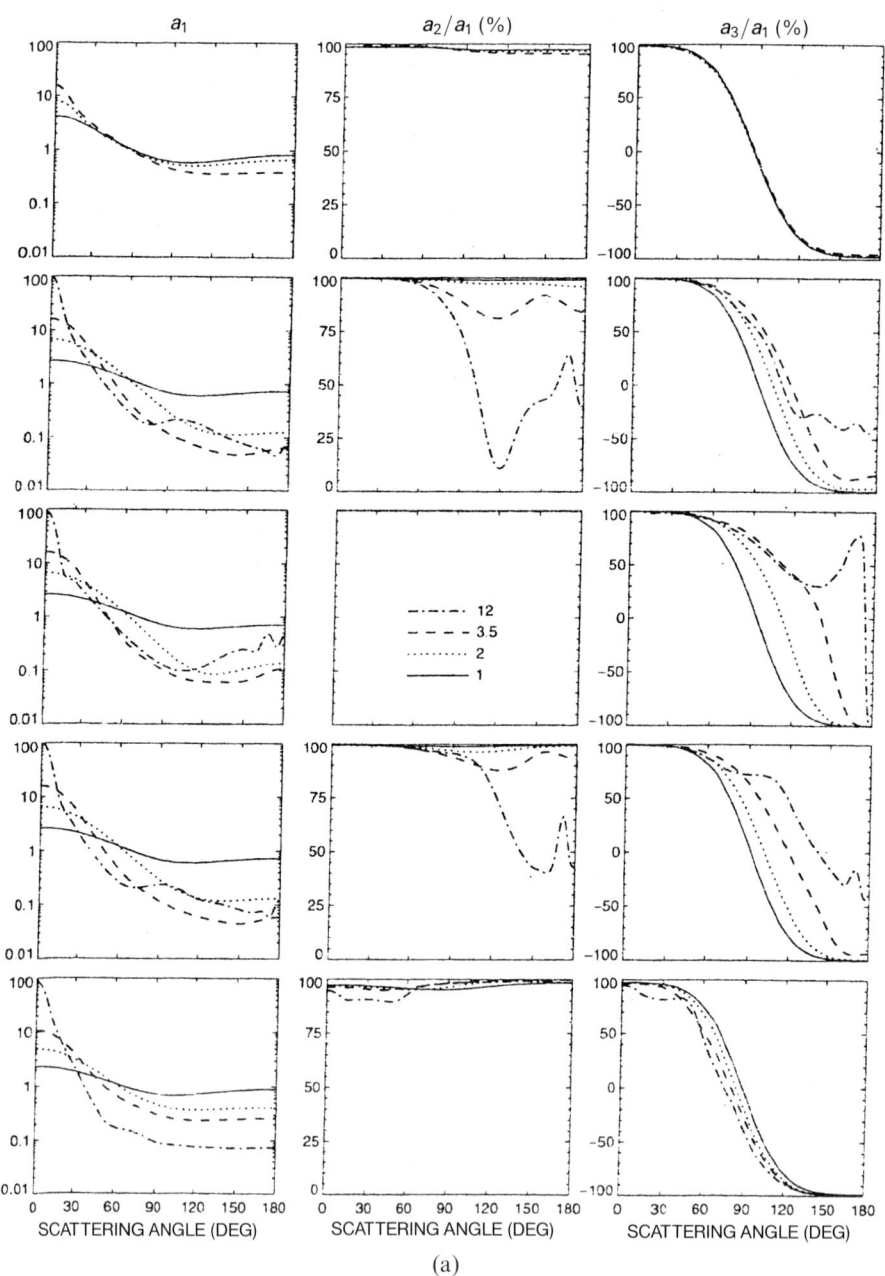

Figure 3.13. (a) The phase function $p(\theta) = a_1$, the matrix element $p_{22} = a_2/a_1$ (in percent), the matrix element $p_{33} = a_3/a_1$ (in percent) versus scattering angle θ for spheres and surface-equivalent randomly oriented spheroids with size parameters ranging from 1 to 12 and axis ratios of 0.05 (first row), 0.5 (second row), 1 (third row), 2 (fourth row), and 20 (fifth row).

Sec. 3.3] Ellipsoidal particles 97

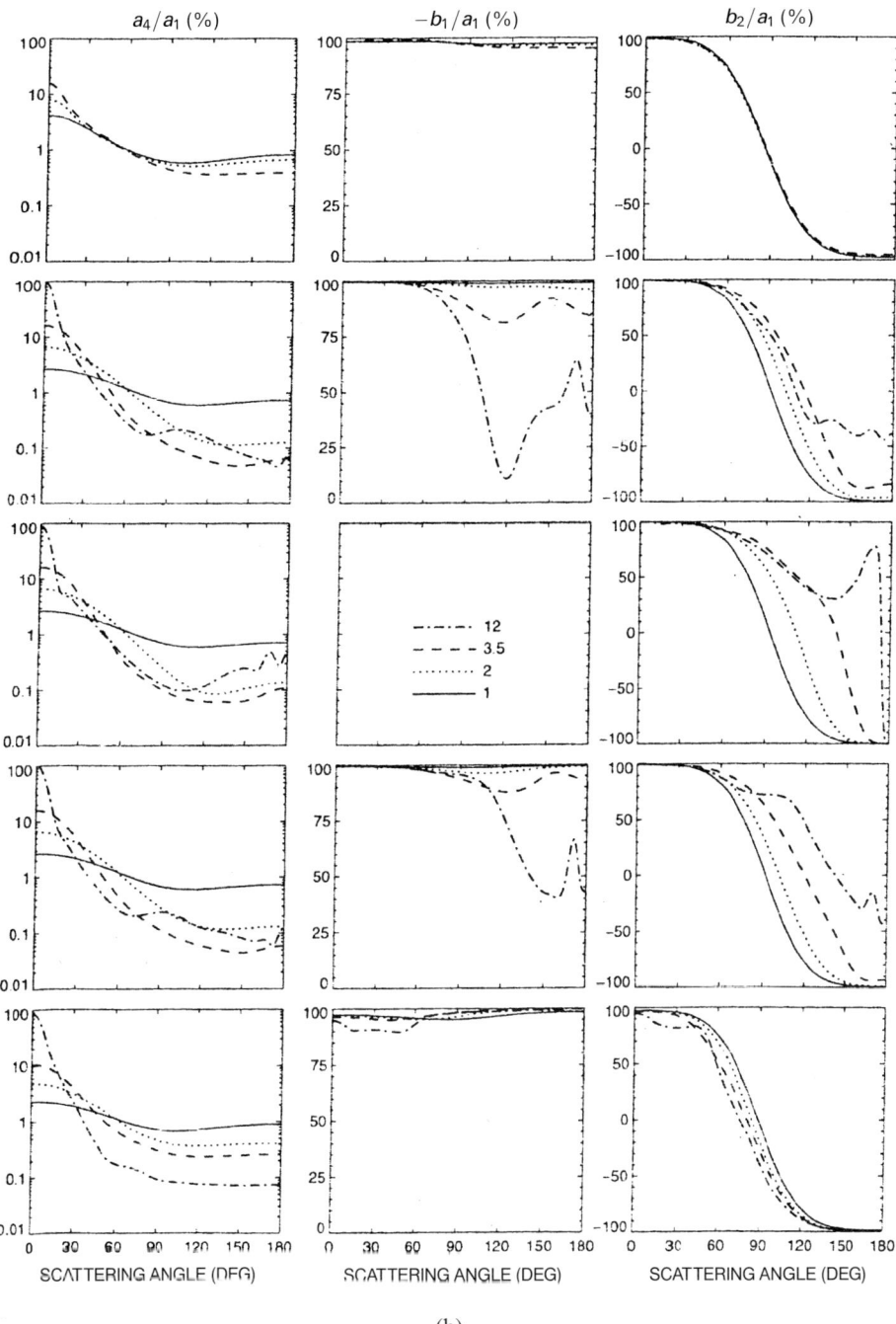

(b)

Figure 3.13 (*cont.*). (b) The same as in (a) but for the matrix elements $p_{44} = a_4/a_1$, $-p_{12} = -b_1/a_1$, and $p_{34} = b_2/a_1$ (from Mishchenko *et al.*, 2002).

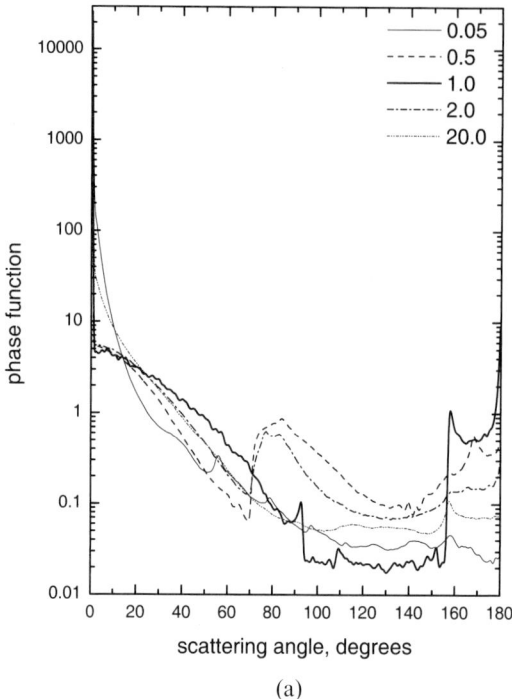
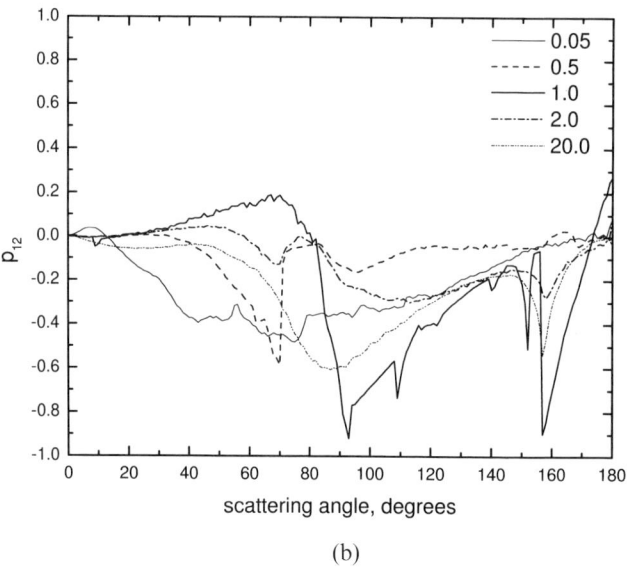

Figure 3.14. (a) The dependence of the phase function $p(\theta)$ on the axis ratio $\xi = 0.05, 0.5, 1.0,$ 2.0, and 20.0 for non-absorbing spheroidal particles at $n = 1.5$, obtained using geometrical optics calculations; (b) the same as in (a) but for p_{12}.

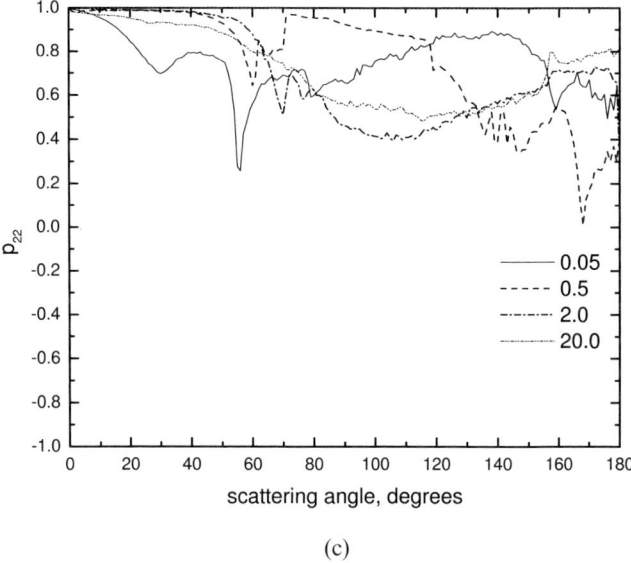

(c)

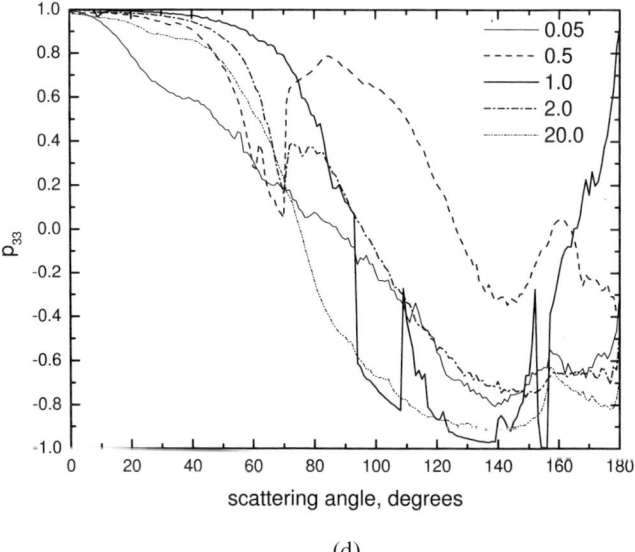

(d)

Figure 3.14 (*cont.*). (c) The same as in (a) but for p_{22}; (d) the same as in (a) but for p_{33}.

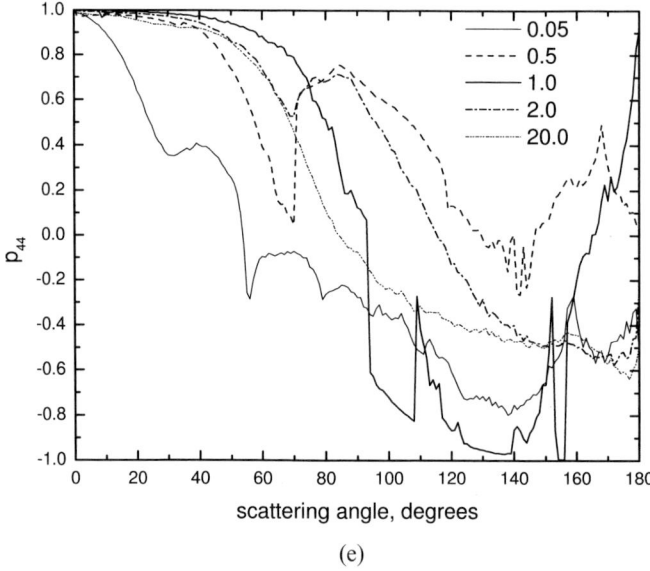

(e)

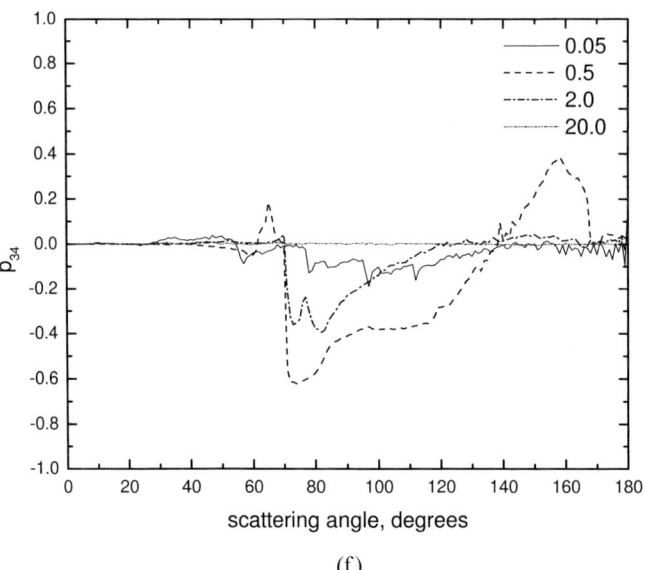

(f)

Figure 3.14 (*cont.*). (e) The same as in (a) but for p_{44}; (f) the same as in (a) but for p_{34}.

on the curves in Figure 3.14(a), which are otherwise smoothed out due to wave optics effects. The geometrical optics technique is combined with the Monte Carlo method in our calculations as described by Macke *et al.* (1996). The small-scale ripple structure is statistical noise from the Monte Carlo method. It can be diminished by taking a larger number of rays or orientations. However, for purposes of a qualitative analysis, we restrict ourselves to 100 particle orientations and 10^4 rays.

First of all we note that the case $\xi = 1$, which corresponds to a spherical particle, gives smallest values of phase functions in the range of scattering angles 90–155° as compared to spheroids. Rainbows are clearly seen for spheroids with values of $\xi = 0.5$ and 2.0. However, they disappear as $\xi \to 0$ and $\xi \to \infty$. The curves become more smooth for these cases. The same effect exists for small particles. The large differences at small scattering angles are due to the different sizes of the particles. However, the effect of size is of no importance outside the diffraction peak due to our assumption that $\lambda/a \to 0, \lambda/b \to 0$, and that there is no absorption. Therefore, results presented in these figures have a universal character in the geometrical optics domain.

The matrix element p_{12} is presented in Figure 3.14(b). As was shown above, the value of $p_l = -p_{12}$ gives the degree of linear polarization p_l of unpolarized incident light by an ellipsoidal particle. We see that spheres have the largest value of $|p_l|$ in the backward hemisphere. Spheroidal particles produce smaller polarization of backscattered light compared to spheres ($\xi = 1$). Generally the opposite is true in the forward hemisphere. Polarization p_l is positive for most angles, which means that the electric vector oscillates predominantly in the plane perpendicular to the scattering plane. The curve $p_l(\theta)$ at $\xi = 20$ is similar in shape to Rayleigh particles but with reduced amplitude.

The element p_{22} is equal to 1 for spheres, and is in the range [0;1] (see Figure 3.14(c)) for spheroidal particles (even as $\xi \to 0$ and $\xi \to \infty$, which is in contrast with the case of small spheroids, see Figure 3.13(a)).

The elements p_{33} and p_{44} coincide for spheres. This is not the case for spheroids (see Figures 3.14(d) and (e)). There is a weak tendency for the elements p_{33} and p_{34} to be close to the case of Rayleigh scattering as $\xi \to 0$ and $\xi \to \infty$. However, this is not so clearly pronounced as in the case of small ellipsoids discussed above.

The matrix element p_{34} is equal to zero for non-absorbing spheres in the geometrical optics approximation (if interference effects are neglected). It is very close to zero for prolate spheroids with $\xi = 20$ (see Figure 3.14(f)). This is due to the fact that such spheroids are close to semi-infinite cylinders, which also have $p_{34} = 0$ in the approximation considered. Generally, p_{34} decreases as $\xi \to 0$ and $\xi \to \infty$. This corresponds to the same finding for small spheroids (see Figure 3.13(b)). Oblate spheroids give larger values of $|p_{34}|$ than prolate ones in general. This will lead to larger values of the degree of circular polarization for light scattered by oblate large spheroids as compared to prolate ones (for the linearly polarized incident light).

The case of spheroidal particles distributed with respect to the axis *a* according to the lognormal law with $ka_{ef} = 7$ and $\Delta_{ef} = 0.2$ is given in Figure 3.15 at $\xi = 5$ and

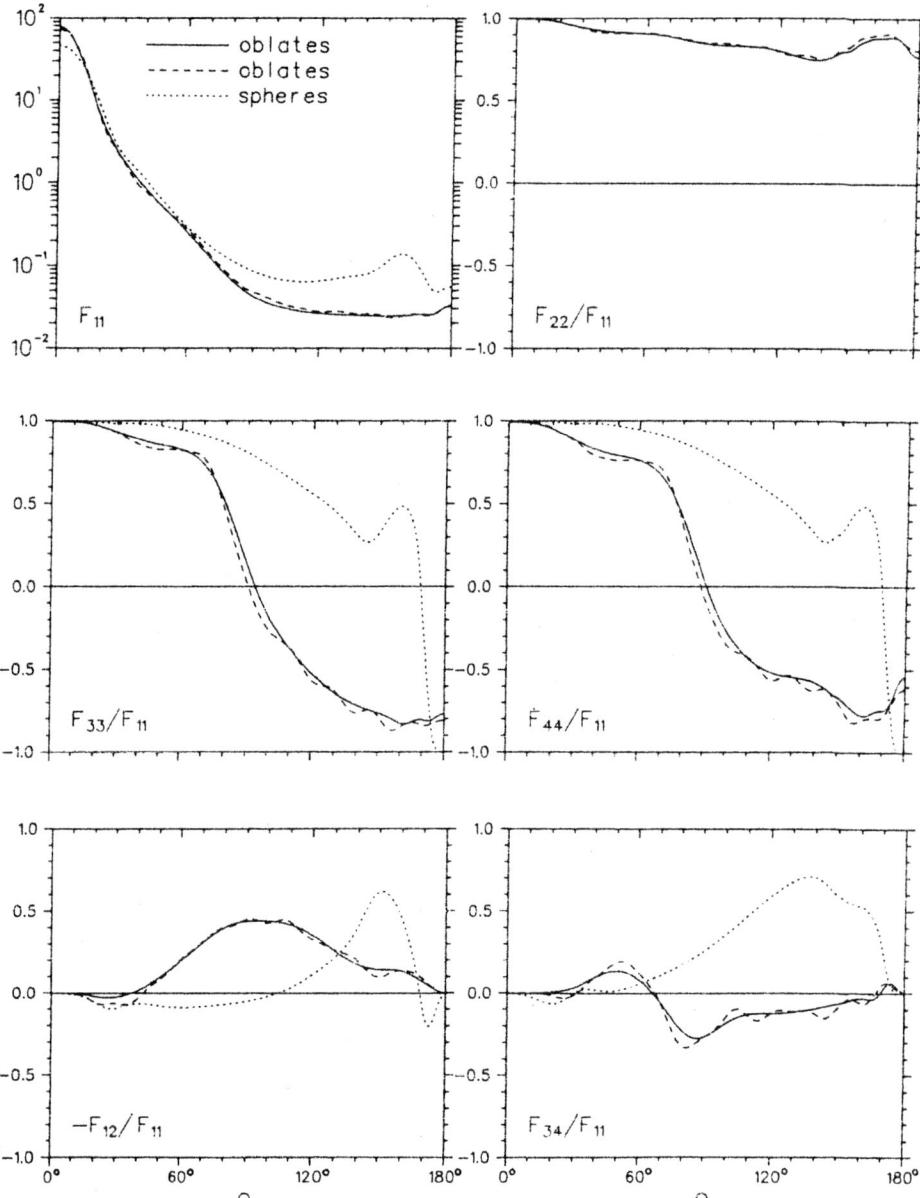

Figure 3.15. The phase function $p(\theta) = F_{11}$, normalized matrix elements $p_{33} = F_{33}/F_{11}$, $-p_{12} = -F_{12}/F_{11}$, $p_{22} = F_{22}/F_{11}$, $p_{44} = F_{44}/F_{11}$, and $p_{34} = -F_{34}/F_{11}$ of randomly oriented oblate spheroids with ratios a/b equal to 5 (solid line) and spheres (points), both having a log-normal size distribution with $x_{ef} = ka_{ef} = 7$ and $\sigma_{ef} = 0.2$. The long-dashed curve gives results for monodispersed randomly oriented spheroids with the size parameter of a volume-equivalent sphere $x = ka = 7$. The refractive index is equal to 1.31 (Stammes, 1989).

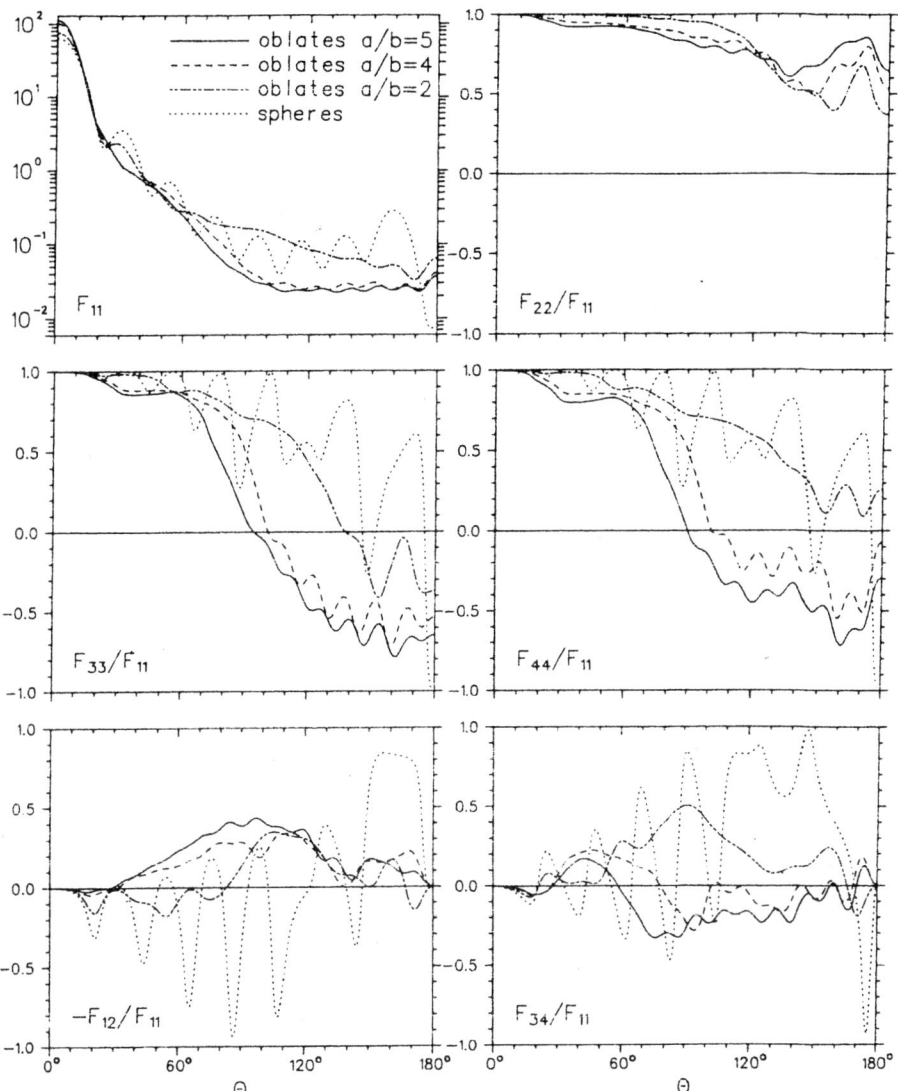

Figure 3.16. The phase function $p(\theta) = F_{11}$, normalized matrix elements $p_{33} = F_{33}/F_{11}$, $-p_{12} = -F_{12}/F_{11}$, $p_{22} = F_{22}/F_{11}$, $p_{44} = F_{44}/F_{11}$, and $p_{34} = -F_{34}/F_{11}$ of spheres with $ka = 8.57$ and volume-equivalent randomly oriented oblate spheroids with axis ratios equal to 5, 4, and 2. The refractive index is equal to 1.31 (Stammes, 1989).

$n = 1.31$. The data for spherical particles and monodispersed spheroids with $ka_v = 7$ are also given for the comparison, where a_v is the radius of the volume-equivalent sphere. We see that randomly oriented spheroids of various sizes can be modelled with a high accuracy by monodispersed spheroids having the same aspect ratio and value of the parameter $a_v = (3/4\pi)\sqrt[3]{V}$ equal to a_{ef}. Here V is the volume of a spheroid (see Table 1.1).

Results for spheroids and volume equivalent spheres differ considerably, however. A comparison of results for spheres and oblate spheroids with $\xi = 2, 4$, and 5 is given in Figure 3.16. We see that the scattering matrix elements for oblate spheroidal particles with $\xi = 2, 4$, and 5 differ. This can be used for particle characterization purposes.

3.4 CYLINDERS

The model of cylindrical particles is also often employed in theoretical studies of light scattering and radiative transfer in various complex disperse media. It has the same number of free parameters as the case of spheroids with the aspect ratio defined as the ratio of the diameter $D = 2a$ to the length L, $\xi = D/L$. Also the shape of the cross section of the cylinder can vary, depending on the application. The most easy case to handle is that of a circular cylinder. The model of hexagonal cylinders allows the halo phenomenon to be described, which often occurs for ice clouds. This produces enhanced scattering at angles of 22 and 46° (at $n = 1.31$). These halo phenomena cannot be described by the ellipsoid particle model.

Extremely extended spheroids have scattering characteristics close to that of infinite cylinders (e.g., Figure 3.17, where only minor differences are present in the phase matrix elements). Therefore, in this case the model of infinite cylinders can be chosen for modelling purposes due to its simplicity.

We compare results of calculations for infinite cylinders, prolate spheroids and spheres in Figure 3.18. It follows that spheroids and cylinders behave in a similar way (with some differences, however [e.g., the element p_{44}]). Results for cylinders with $\xi = 1, 3$, and 5 and $\xi \to \infty$ are given in Figure 3.19. They were obtained at $ka = 2.75$ and $m = 1.31$. This corresponds to ice in the visible range of the electromagnetic spectrum.

It is interesting to note that the imaginary part of the refractive index does not influence the matrix elements (compare Figures 3.18(a) and 3.18(b)). This is mostly due to the small size of the particles. For larger particles, the value of χ is of importance as was discussed above (see Figure 3.3(a)).

It follows from Figure 3.19 that the angular behaviour of elements p_{33} and p_{44} is similar. The value of p_{22} is larger than 0.5. Also the degree of polarization p_l is mostly positive with a shift to larger angles as compared to Rayleigh particles. The value of $|p_{34}|$ is smaller than 0.5. The results for $\xi = 5$ are close to that at $\xi \to \infty$. However, they do not coincide. The value of ξ should be larger than 10 to have the model of infinite cylinders valid.

The results of computations for large ($kD \to \infty$, and $kL \to \infty$) hexagonal cylinders at $n = 1.5$ and $\xi = D/L = 1$ are given in Figures 3.20 and 3.21. The value of D is the distance between sides of a hexagonal cross section in this case. We see that the degree of polarization $p_l = -p_{12}$ is small. Note also that there is a

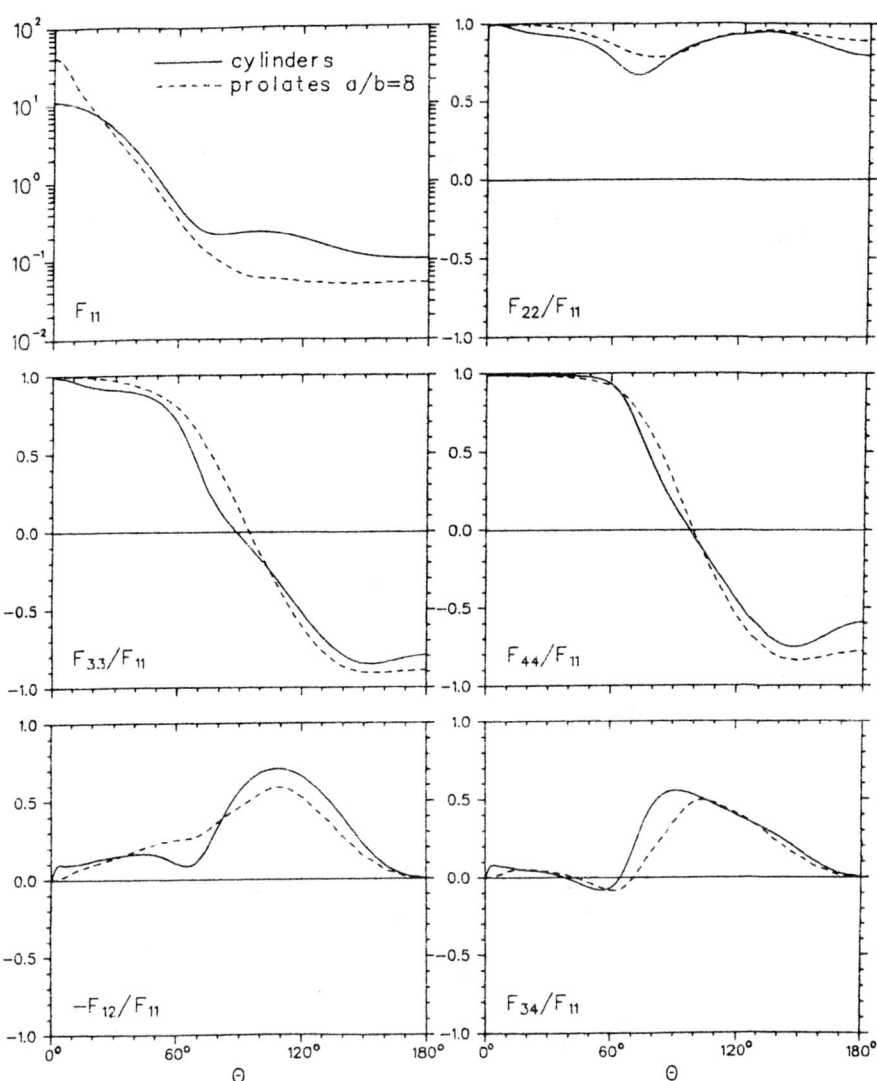

Figure 3.17. The phase function $p(\theta) = F_{11}$, normalized matrix elements $p_{33} = F_{33}/F_{11}$, $-p_{12} = -F_{12}/F_{11}$, $p_{22} = F_{22}/F_{11}$, $p_{44} = F_{44}/F_{11}$, and $p_{34} = -F_{34}/F_{11}$ of randomly oriented infinite cylinders with the size parameter 2.5 and randomly oriented prolate spheroids with the aspect ratio ξ^{-1} equal to 8 and the size parameter correspondent to the smallest axis being equal to 2.5. The refractive index is equal to 1.31 (Stammes, 1989).

halo close to $38°$ ($n = 1.5$). Generally, the most pronounced halo for hexagonal cylinders appears at the scattering angle (Tricker, 1970):

$$\theta_h = 2\arcsin\left(\frac{n}{2}\right) - \frac{\pi}{3}.$$

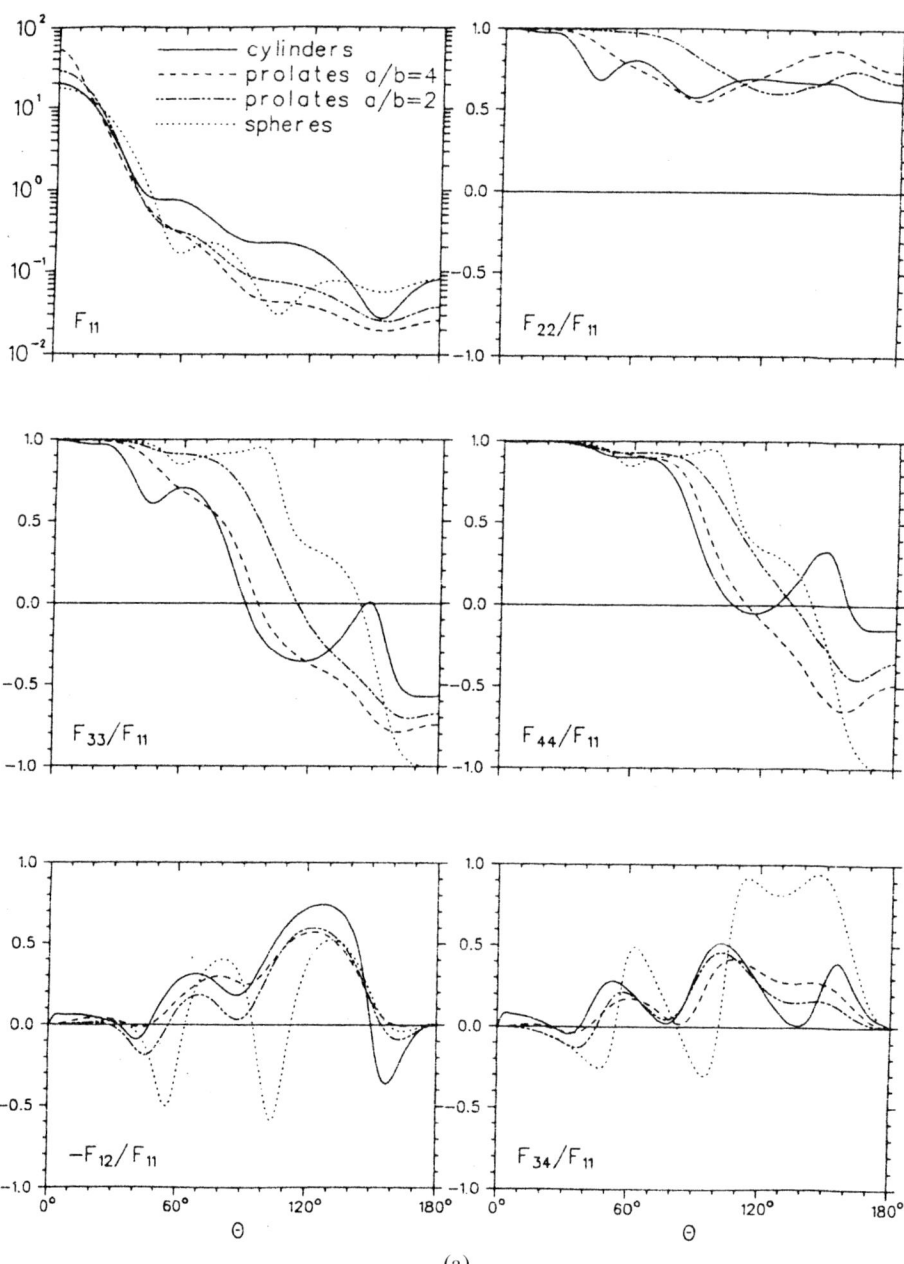

Figure 3.18. (a) The phase function $p(\theta) = F_{11}$, normalized matrix elements $p_{33} = F_{33}/F_{11}$, $-p_{12} = -F_{12}/F_{11}$, $p_{22} = F_{22}/F_{11}$, $p_{44} = F_{44}/F_{11}$, and $p_{34} = F_{34}/F_{11}$ of randomly oriented infinite cylinders and spheres with the size parameter 4, randomly oriented prolate spheroids with the aspect ratio 4 and the size parameter correspondent to the smallest axis being equal to 4. The refractive index is equal to 1.31.

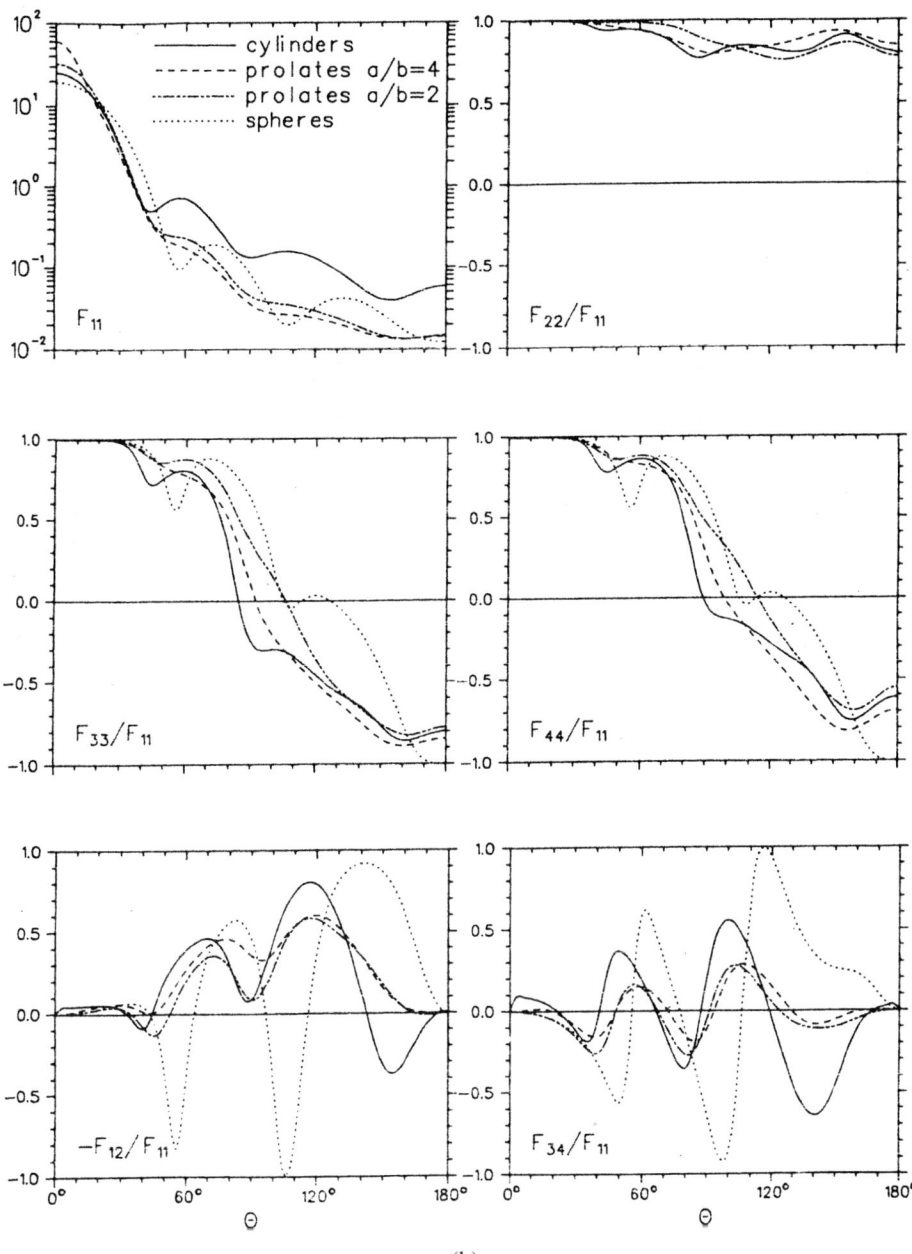

Figure 3.18 (*cont*). (b) the same as in (a) but for $m = 1.31 - 0.1i$ (from Stammes, 1989).

108 Local optical characteristics [Ch. 3

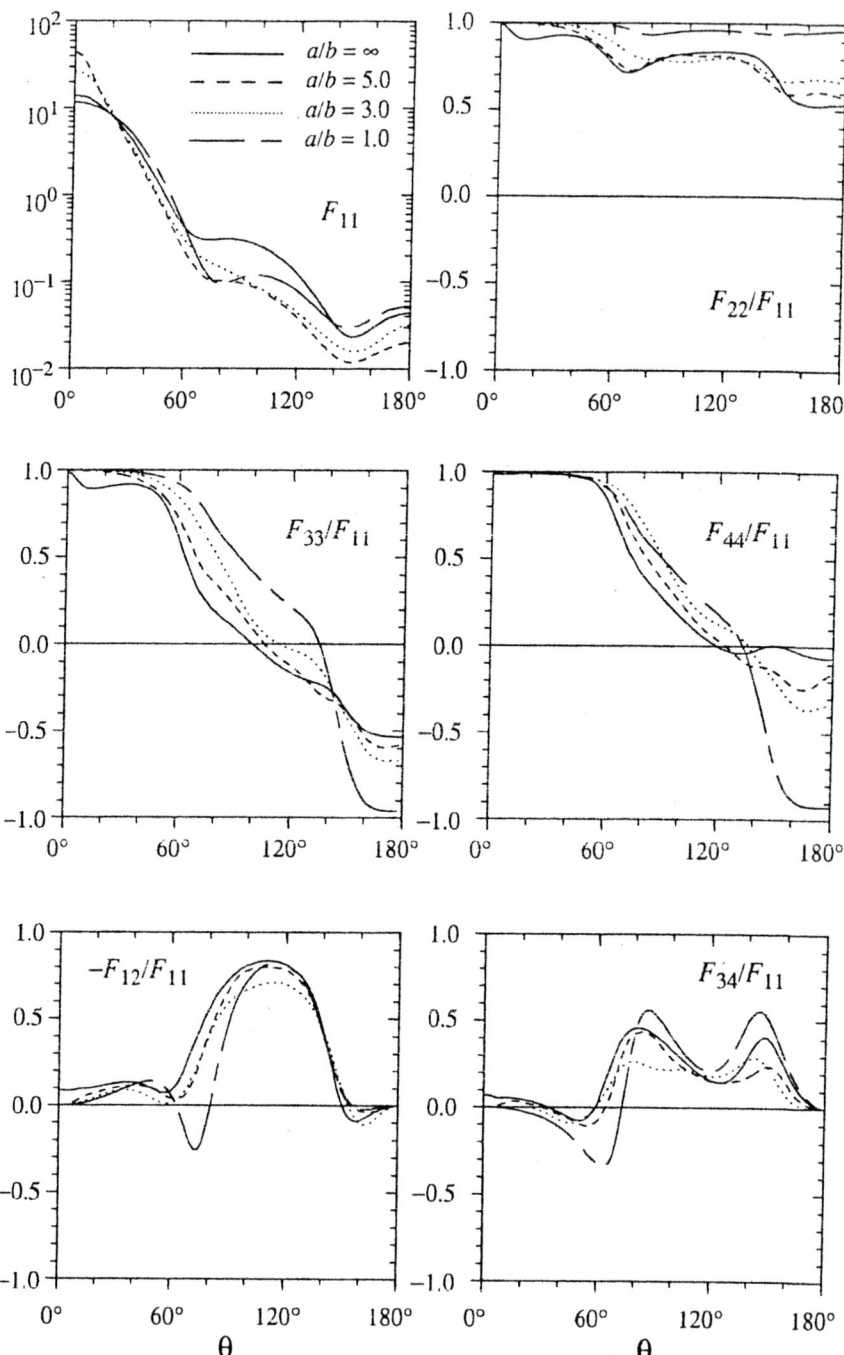

Figure 3.19. The same as in Figure 3.16 but for randomly oriented circular cylinders with the size parameter 2.75 and various ratios aspect ratios $\xi^{-1} = 1$, 3, and 5 (from Kuik, 1992).

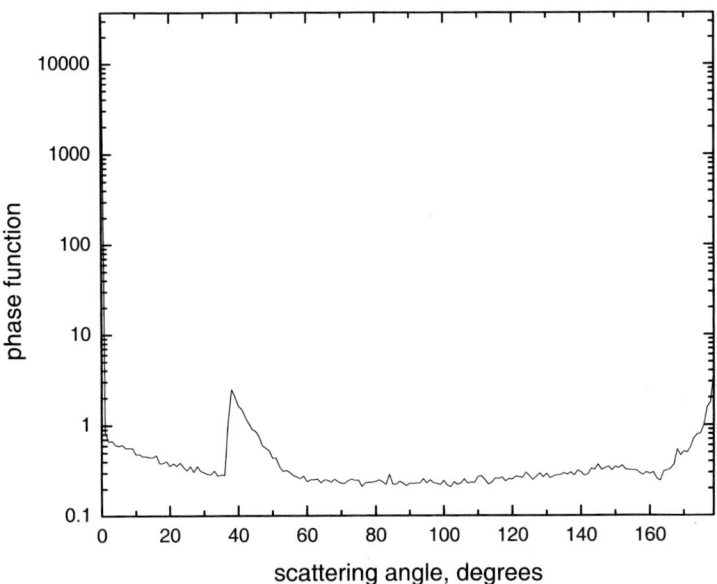

Figure 3.20. The phase function of randomly oriented non-absorbing finite hexagonal cylinders obtained in the framework of the geometrical optics approximation at the refractive index equal to 1.5. The aspect ratio is equal to 1.

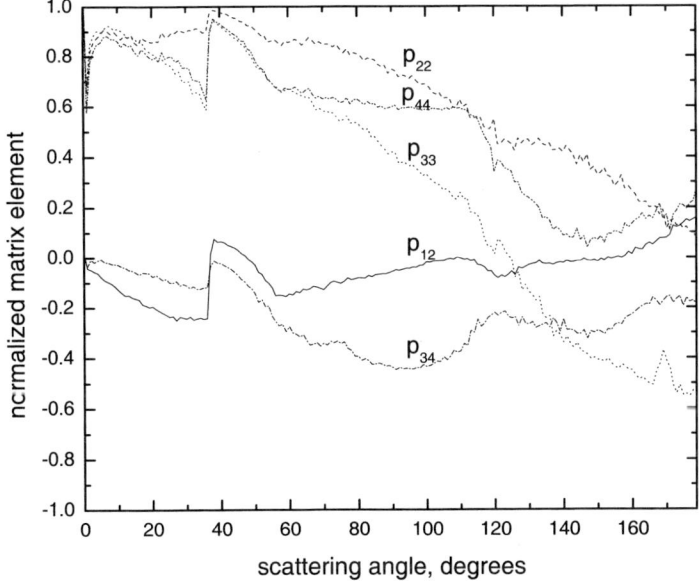

Figure 3.21. The same as in Figure 3.20 but for the normalized phase matrix elements.

Table 3.7. The dependence of the halo angle for the hexagonal cylinder on the real part of the refractive index n.

n	1.0	1.1	1.2	1.31	1.4	1.5	1.6	1.7	1.8
θ_h(degrees)	0.0	6.73	13.74	21.84	28.85	37.20	46.26	56.42	68.32

The dependence of θ_h on the real part of the refractive index is given in Table 3.7. Note that for cubes (Tricker, 1970):

$$\theta_h = 2\arcsin\left(\frac{n}{\sqrt{2}}\right) - \frac{\pi}{2}.$$

Therefore, the halo phenomenon and its position can be used for shape and refractive index characterization.

3.5 IRREGULARLY SHAPED PARTICLES

3.5.1 Koch fractals

Natural media are usually composed of spherical particles (e.g., liquid aerosols and water clouds) or irregularly shaped particles (e.g., solid aerosols and ice clouds). The scattering of light by spherical particles can be described with high accuracy by Mie theory as discussed above. For irregularly shaped particles, the combination of simple regular shapes with different weights in a total sum is often used (Liou, 2002).

Another approach to the modelling of the optical response of irregularly shaped particles is based on a stochastic single particle model. This particle should have similar light scattering characteristics to the ensemble of irregularly shaped particles under investigation. Then, instead of varying the weights, we can vary the parameters of this stochastic particle.

We consider here, in some detail, a particular model of a stochastic particle, developed by Macke (1994) and Macke et al. (1996). This is based on the notion of the fractal (Mandelbrot, 1977). The fractal particle is constructed in the following way. The initial shape is a tetrahedron. A part of its triangular surface is replaced by a reduced version of the same tetrahedron. The resulting body is called the first generation of the triadic Koch fractal. Repetition of this procedure at the smaller triangles leads to higher generations. We will consider here the Koch fractals of the second generation. Optical properties of Koch fractals of the second and higher generations differ insignificantly (Macke et al., 1996).

The interaction of light with a large fractal particle is studied in the framework of geometrical optics, using the Monte Carlo ray-tracing technique. Results are averaged with respect to the particle random orientation. The peak in the forward direction is described in the framework of Fraunhofer diffraction theory (Born and Wolf, 1999).

We also assume that all planes of a fractal particle have a roughness. The roughness is modelled in the following way (Macke et al., 1996). For each reflection–refraction event the normal of the crystal surface is tilted randomly around its original direction. The azimuth tilt angle is chosen randomly with equal distribution from the interval $[0, 2\pi]$. The zenith tilt angle is chosen from the range $[0, \pi s/2]$, where s is the roughness parameter. We assume that $s = 0.3$ in this study. Larger values of s influence the result insignificantly.

We assume that the wavelength is 0.5 µm and the refractive index n varies in the range 1.1–1.5, which corresponds to most natural disperse media in the visible. The length D of the initial tetrahedron's side is equal to 100 µm.

The particle is assumed to be non-absorbing. This means that its normalized phase matrix does not depend on size in the geometrical optics approximation (except at the small-scattering region, where the Fraunhofer diffraction takes place). This makes the model inflexible as far as real experimental data (which can vary for different media) are concerned. On the other hand there is an advantage in this. Namely, we have a fixed phase matrix for chaotic scattering and measured matrices for various media can be checked against this simple case of a single fractal particle without introducing any fitting parameters (if the refractive index is known).

Let us now discuss results of numerical calculations. The first element of the phase matrix P_{11} (or the phase function) is given in Figure 3.22(a) at various values of the refractive index n for the case of a fractal particle, as described above. Fractal particles with refractive index $n = 1.5$ have a phase function which is almost isotropic in the backward hemisphere. Such phase functions were experimentally measured for dust aerosols (von Hoyningen-Huene and Posse, 1997; Volten, 2001; Volten et al., 2001).

Phase functions in Figure 3.22(a) are almost featureless and have smaller values for smaller n at scattering angles larger than $\pi/4$. For smaller angles ($\theta < \pi/4$), the opposite is true. Note that the oscillations in Figure 3.22(a) (and in figures, which follow) are due to statistical noise in the Monte Carlo code.

It should be emphasized that the phase function for systems of fractal particles does not depend on the refractive index of particles around $\theta \approx 45°$ (see Figure 3.22(a)). This may explain the existence of a cross-point at this angle for experimentally measured oceanic phase functions (Tyler, 1961; Zaneveld and Pack, 1973). It is known that oceanic water contains a great portion of irregularly shaped particles, having various refractive indices (Shifrin, 1988; Sid'ko et al., 1990; Kokhanovsky, 2001). Also, oceanic phase functions are featureless, similar to those given in Figure 3.22(a) (e.g., the case $n = 1.1$).

The case $n = 1.3$ in Figure 2.22(a) corresponds to ice fractal particles in the visible and reproduce well the phase functions of crystalline clouds (Macke et al., 1996). The case $n = 1.5$ roughly corresponds to dust particles (Volten, 2001; Volten et al., 2001).

The asymmetry parameter g was found using the data given in Figure 3.22(a) and also calculations at other values of n (see Table 3.8).

This parameter can be presented in the following general form for large non-absorbing particles (Kokhanovsky, 2001):

$$g = \frac{g_d + g_0}{2},$$

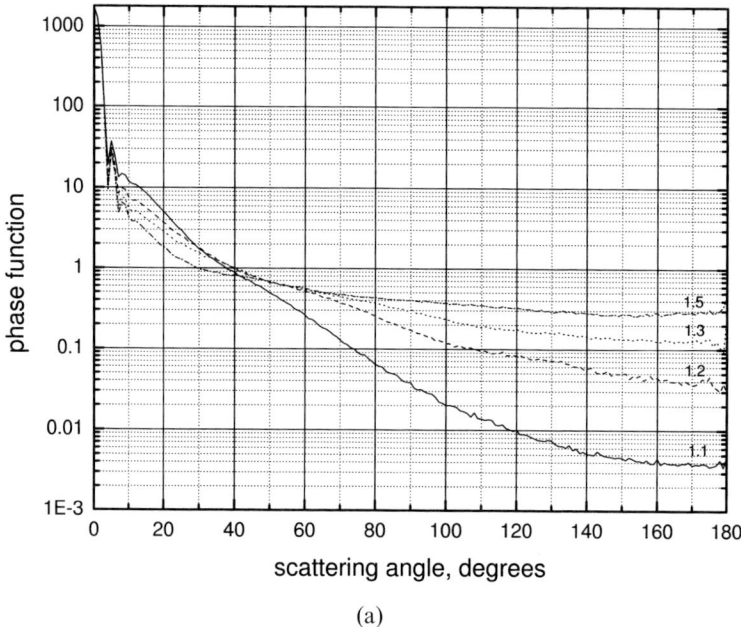

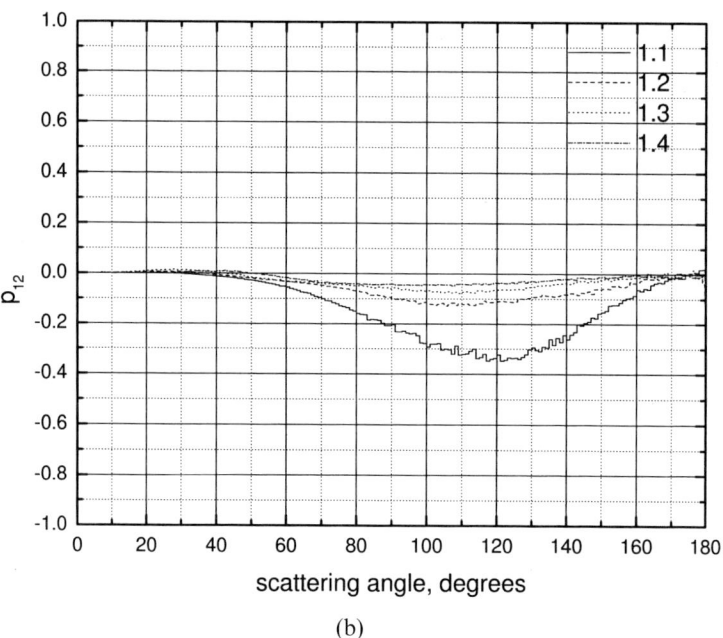

Figure 3.22. (a) The phase function of fractals for various values of the real part of the refractive index; (b) the same as in (a) but for the element p_{12}.

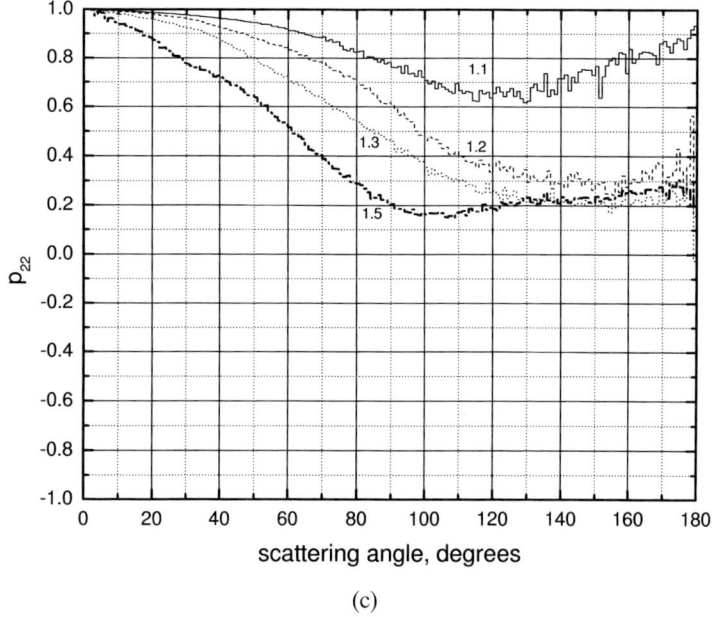

(c)

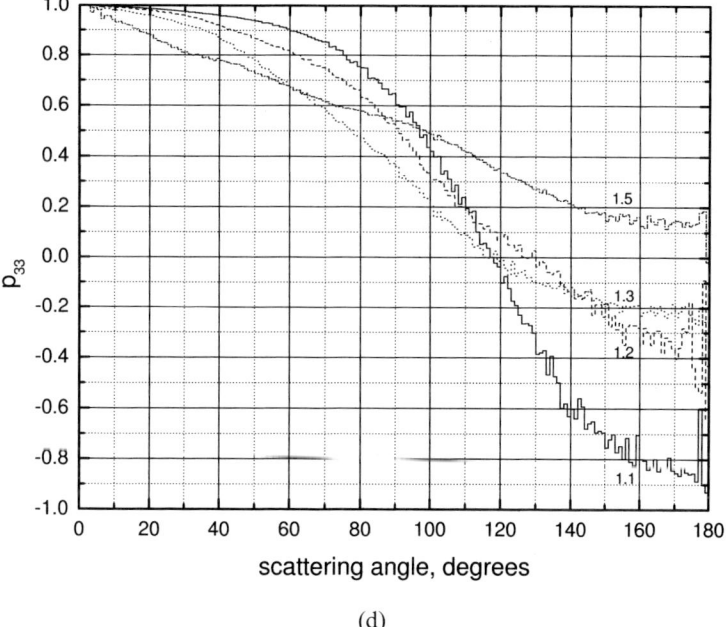

(d)

Figure 3.22 (*cont.*). (c) The same as in (a) but for the element p_{22}; (d) the same as in (a) but for the element p_{33}.

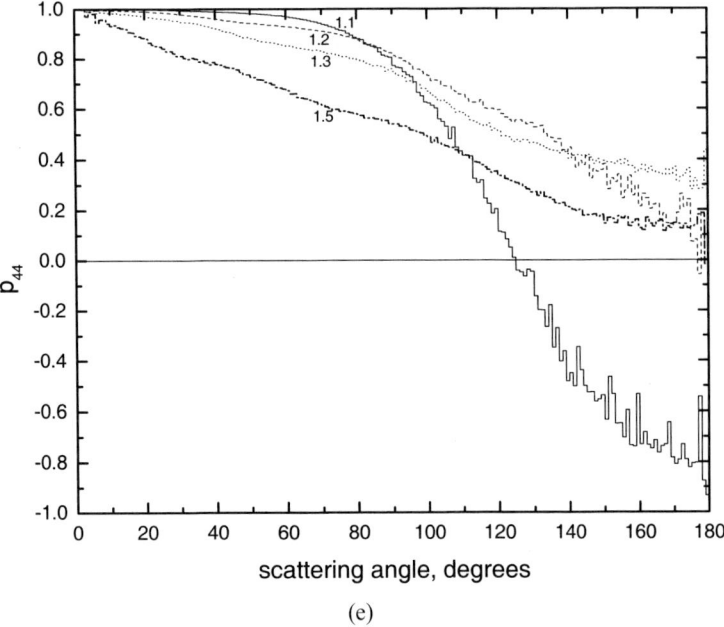

(e)

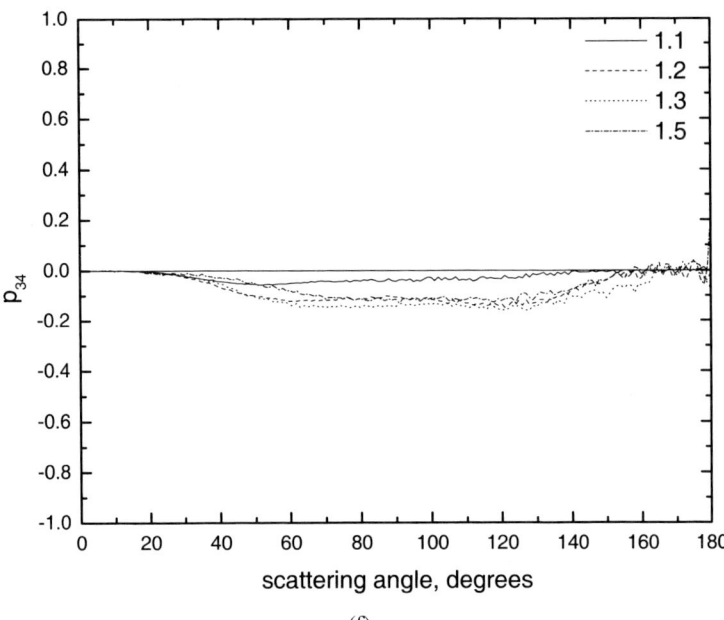

(f)

Figure 3.22 (*cont.*). (e) The same as in (a) but for the element p_{44}; (f) the same as in (a) but for the element p_{34}.

Table 3.8. The asymmetry parameters g_0 and g for various n.

n	g_0	g
1.0	1.0	1.0
1.1	0.86	0.93
1.2	0.68	0.84
1.31	0.48	0.74
1.4	0.39	0.69
1.5	0.30	0.65
1.6	0.23	0.61
1.7	0.18	0.59
1.8	0.15	0.57
1.9	0.12	0.56
2.0	0.10	0.55
∞	0.0	0.50

where $g_d \approx 1$ is the asymmetry parameter related to the diffraction and g_0 is the asymmetry parameter related to the geometrical optics scattering process (see Table 3.8). Light scattering by a fractal particle leads to an isotropic distribution of photon scattering directions at large n (see Figure 3.22(a)). Thus, we get $g_0 \to 0$ as $n \to \infty$. This is confirmed by data given in Table 3.8. Note that for ice crystals in the visible spectrum ($n = 1.31$) we obtain $g = 0.74$, which coincides with the experimental result obtained from measurements performed by Garett et al. (2001) in natural crystalline clouds.

The asymmetry parameter g is approximately 0.93 at $n = 1.1$, which is typical for oceanic hydrosols with particles having a refractive index closed to that of water (Shifrin, 1988). Clearly, we get: $g_0 \to 1$ as $n \to 1$ (see Table 3.8).

The following parameterization holds with an error less than 5% at $n \leq 1.9$:

$$g_0 = 0.1 + 1.2\exp(-\mathbb{C}(n - n_0)^2),$$

where $\mathbb{C} = 2.65$ and $n_0 = 0.68$. This approximate equation for the asymmetry parameter can be used for modelling of radiative transfer in ice clouds using selected approximations (Kokhanovsky, 2001) which do not require information on the phase function.

Let us consider the normalized phase matrix elements $p_{ij} = P_{ij}/P_{11}$. We start with the element p_{12}. The degree of polarization of initially unpolarized light after scattering of light by a fractal particle is equal to $-p_{12}$. The element p_{12} is given in Figure 3.22(b) for various n. It is negative for most angles, which means that oscillations of an electric vector are predominantly in the plane perpendicular to the scattering plane. Generally, the curves in Figure 3.22(b) are similar to the case of Rayleigh scattering but with strongly reduced values (e.g., it is only 5% polarization at maximum at $n = 1.5$ as opposite to the Rayleigh case, where the maximal polarization is equal to 100% at $\theta = 90°$). Such a similarity also follows from experiments (Volten et al., 2001). Note also the shift of the maximum polarization angle $\theta_{\max}$ to larger angles as compared to Rayleigh scattering. Very small values of the degree of polarization and curves similar to that in Figure 3.22(b) were found experimentally for oceanic water (Beardsley, 1968), ice clouds (Dugin and Mirumyants, 1976) and Saharan sand particles (Volten et al., 2001).

Generally, the degree of polarization decreases with n. Larger values of n lead to a larger degree of randomization of photon polarization states after a scattering event.

We see that fractal particles do not produce a strong polarization of incident unpolarized light. This is a general feature of disperse media with irregularly shaped particles (Volten et al., 2001).

The element p_{22} is given in Figure 3.22(c). It is equal to one for spherical particles, so its difference from unity is the measure of the particle's non-sphericity. We see that deviations from the spherical particle case ($p_{22} = 1$) grow with n. This result is easily understood. For optically soft particles there is almost no shape dependency of the light scattering characteristics (Sid'ko et al., 1990).

Note also broad minima in Figure 3.22(c) at $n = 1.1$ and $n = 1.5$. Such minima were observed by Voss and Fry (1984) and Volten (2001) for oceanic water and quartz aerosol respectively.

The element p_{33} is given in Figure 3.22(d). We see that curves $p_{33}(\theta)$ are close to each other around a scattering angle $20°$, where the value of p_{33} is small. The absolute values of p_{33} decrease with the refractive index in the forward hemisphere.

The element p_{34} is given in Figure 3.22(e). It takes small negative values. This element describes the efficiency of the transformation of linear polarized light to a circularly polarized mode for a given scattering process. Such a transformation for fractal particles has low probability as seen from Figure 3.22(e). Generally, p_{34} decreases with n.

The element p_{44} is given in Figure 3.22(f). Note that the case $n = 1.1$ differs considerably from the case $n \geq 1.2$. The value of p_{44} describes the reduction of the degree of circular polarization for an initially completely circularly polarized incident light beam.

Angular dependencies of the degree of polarization $q(\theta)$ and the cross-polarization ratio $r(\theta) = (1 - p_{22})/(1 - 2p_{12} + p_{22})$ are given in Figures 3.23(a) and (b) at $n = 1.31$ and $n = 1.5 - i0.005$, which roughly correspond to ice clouds and dust aerosols respectively. The value of r is equal to the ratio of scattered light intensities I_1/I_2. Here I_1 corresponds to the case when a disperse medium is placed between vertical (input) and horizontal (output) polarizers. I_2 corresponds to the case of a disperse medium placed between vertical polarizers. It follows from Figure 3.23(b) that the cross polarization r by ice crystals is larger than that for dust particles in the backward hemisphere. This is mostly due to the effect of light absorption by aerosols, which diminishes the influence of the particle shape on backscattering. The value of q (see Figure 3.23(a)) is generally lower for ice crystals in the backward hemisphere. It remains smaller than 0.3 in this angular range for ice crystals. Remember that $q = 1$ for identical spheres. We see that the backscattering region is potentially informative for discrimination between different particulate media (Card and Jones, 1999).

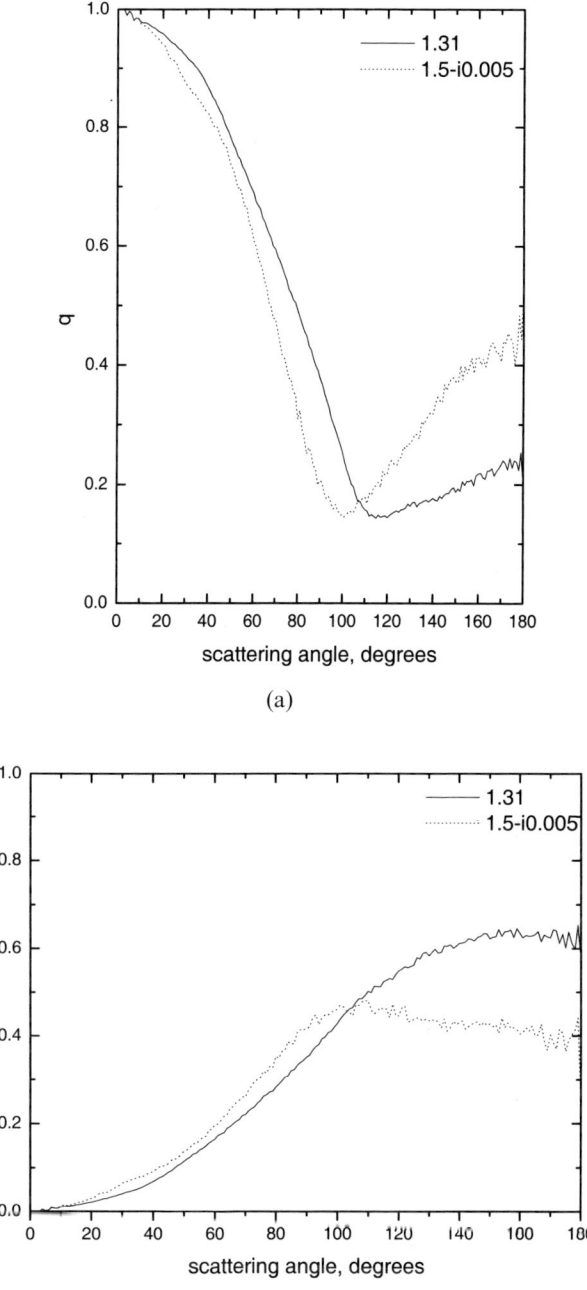

Figure 3.23. (a) The dependence of the parameter q on the scattering angle for ice clouds ($n = 1.31$) and dust aerosols ($m = 1.5 - i0.005$), calculated in the framework of the fractal particle model at $D = 100\,\mu\text{m}$ and $\lambda = 0.5\,\mu\text{m}$; (b) the same as in (a) but for r.

3.5.2 Experimental results

Ice clouds

Let us compare results obtained from the fractal particle model (FPM) with experimental data, starting with the case of ice clouds. Ice clouds are composed of particles of various shapes. Their phase matrices were measured by Dugin and Mirumyants (1976). Results of comparison of their experiments with phase matrix elements at $n = 1.3$, which is close to the refractive index of ice in the visible, where the measurements were performed, are given in Figure 3.24. Let us analyse them.

We see that the measured and calculated functions $p_{12}(\theta)$ have similar shapes, producing a broad negative minimum at side scattering angles. The absolute values, given by the experiment and the theory differ, however. This comes as no surprise. Indeed, the degree of polarization in measurements is obtained as a ratio $(i_1 - i_2)/(i_1 + i_2)$, where i_1 and i_2 are scattered intensities for two orthogonal polarizations. They are small and measured with an estimated error of up to 20%. This means that their differences, which are small numbers, can have errors exceeding 40% or more which are not acceptable for any comparisons with the theory. Also, the crosses represent averaging over 5 experiments. Actual experimental data are scattered almost uniformly in the range $p_{12} \in [-0.2, 0.0]$ (Dugin and Mirumyants, 1976).

The accuracy of the FPM for elements p_{33} and p_{34} is quite high, especially taking into account that no fitting parameters have been used. The sign of p_{34} was different in the experiment due to a different definition of the Stokes vector p_{34} by Dugin and Mirumyants (1976). So the experimental data for p_{34} have been multiplied by $j = -1$ and plotted in Figure 3.24(a).

The differences between the calculated and measured values of p_{44} and p_{22} are quite high. However, we note again that crosses (for p_{44}) and points (for p_{22}) present only averaged data. Real experimental data are scattered around these averaged curves. The theoretical curves for both p_{44} and p_{22} are practically inside the experimental data scattering area, given by Dugin and Mirumyants (1976). Interestingly, measured values of p_{22} are closer to calculated values of p_{44} and vice versa. We do not have an explanation for such a peculiarity.

Overall, the FPM describes (at least qualitatively) the main features of polarized light scattering by ice clouds. It can be used as input to the vector radiative transfer equation for studies of polarized radiative transfer in realistic crystalline cloudy media. Spherical particle models cannot be used in this case. Interestingly, the FPM has no fitting parameters for crystalline clouds in the visible region of the electromagnetic spectrum.

Takano and Liou (1995) compared the experimental results of Dugin and Mirumyants (1976) with theoretical ray-tracing calculations, assuming the following forms of particles: solid plates and dendrites. They stated that the case of dendrites is closer to experimental data. Although the theoretical data, given by Takano and Liou (1995) are generally closer to the experimental results than the FPM calculations presented above, their theoretical curves have special features,

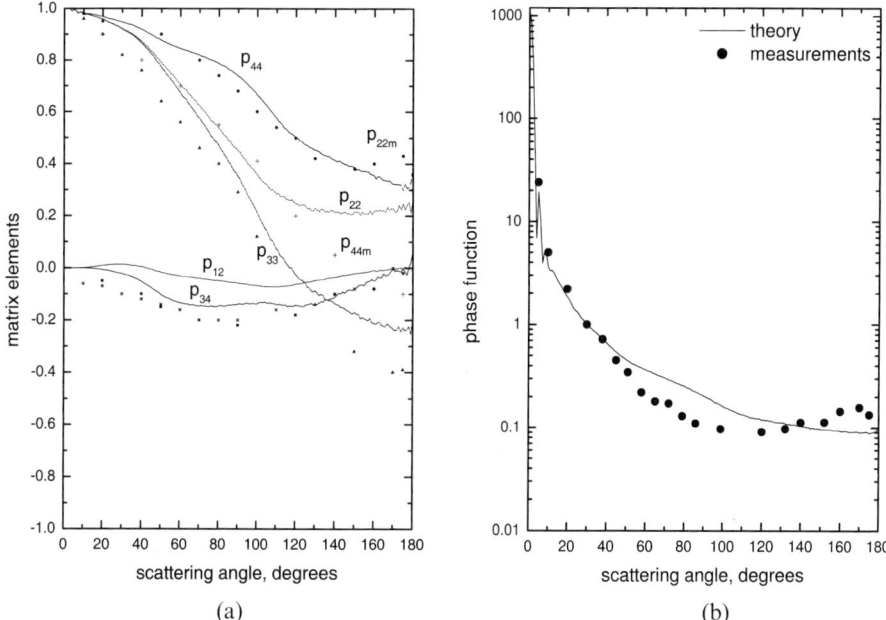

Figure 3.24. (a) The comparison of calculations (lines) and measurements (symbols, see paper by Dugin and Mirumyants, 1976) of the normalized phase matrix elements. Points and crosses give the measured values p_{22m} and p_{44m}, respectively. Triangles, squares, and stars correspond to the measured values p_{33m}, p_{12m}, and p_{34m}, respectively; (b) the phase function of an artificial crystalline cloud, obtained from both measurements (Barkey and Liou, 2001) and the fractal particle model.

which were absent both in the experiment and in the results obtained with the FPM. Thus, the FPM captures the main effects of irregularity on cloud polarization characteristics in a correct way as compared to modelling ice cloud particles by plates or dendrites. N-particle models (Liou, 2002) can provide a better overall accuracy because they have more fitting parameters. Fitting parameters are virtually absent in the FPM, however. This is a main advantage of the model considered here.

In conclusion, we present a comparison of the phase function derived in the framework of the FPM with experimental measurements, performed by Barkey and Liou (2001) (see Figure 3.24(b)). The theoretical results and experimental data corresponded to each other quite well. The difference can be attributed to the fact that a lot of small crystalline particles (and, possibly, spherical droplets) were present in the chamber during the experiment. Measurements for natural ice clouds give a featureless phase function in the backward scattering region (Baran et al., 2002), which corresponds well with results obtained from the FPM. Note that the statistical analysis of cloud phase functions has been performed by Jourdan et al. (2003).

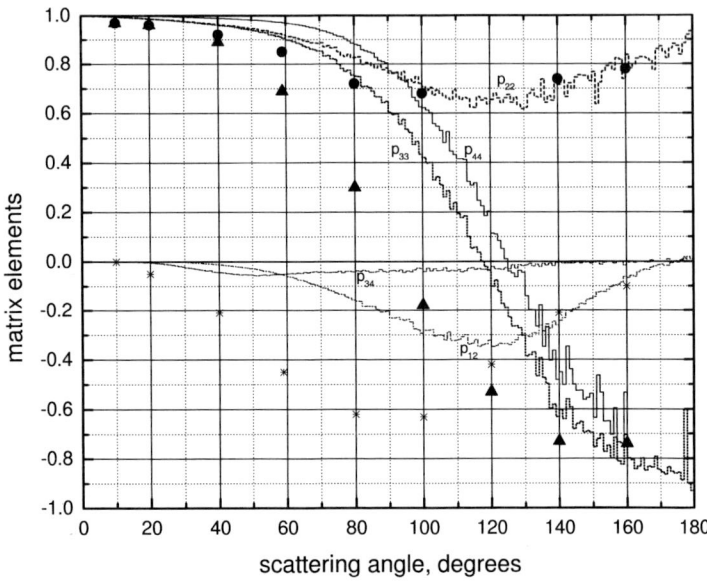

Figure 3.25. Normalized matrix elements, obtained from measurements (circles – p_{22}, triangles – p_{33}, stars – p_{12}) (from Voss and Fry, 1984), and calculations at $n = 1.1$ (lines).

Oceanic water

Oceanic water is another example of a medium with irregularly shaped particles (Shifrin, 1988; Kokhanovsky, 2001). The modelling is much more complex in this case. First of all, oceanic particles have different refractive indices (usually in the range 1.05–1.2), which complicates the analysis as compared to the case of a well known refractive index for ice. Also, both large and small size particles play a role. Moreover, the Rayleigh scattering and *a priori* unknown light absorption by particles and water itself should be accounted for in the calculation of the phase matrix. For the sake of simplicity, however, we neglect all these features and present the results of a comparison of the averaged phase matrices of oceanic water (Voss and Fry, 1984) with results of calculations at $n = 1.1$ in the framework of the FPM shown in Figure 3.25.

We have better overall agreement between the experiment and theory at scattering angles larger than 120°. For smaller angles however, the difference is considerable (especially for elements p_{12} and p_{33}). Voss and Fry (1984) found that $p_{33} \approx p_{44}$ and $p_{34} \approx 0$ for oceanic water. This is confirmed by the FPM (see Figure 3.25).

Note that small values of p_{12} (similar to those which follow from the FPM) have also been reported (Beardsley, 1968). General angular dependencies $p_{12}(\theta)$ and $p_{33}(\theta)$, obtained from the fractal model at $n = 1.1$ and from the experiment, are similar but the actual values differ. This could be due to the influence of Rayleigh scattering by water and other factors, which have been mentioned above. The FPM data presented in Figure 3.25 can be used as input parameters for the theoretical

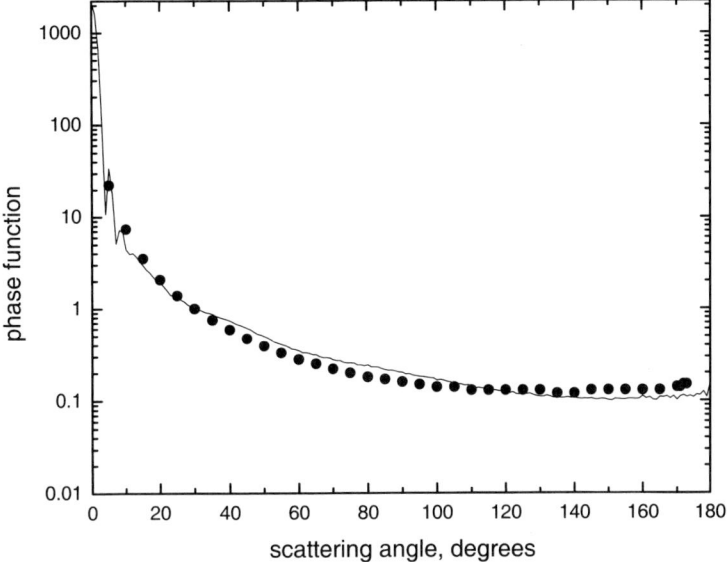

Figure 3.28. The phase function, normalized at the scattering angle of 30°, obtained from measurements (symbols from Volten et al., 2001) and calculated using the fractal particle model at the complex refractive index $1.4-0.005i$ with $D = 100\,\mu m$ and $\lambda = 0.5\,\mu m$.

the element p_{44} compared to the measurements (see Figure 13 in Volten et al. (2001)). This means that both FPM and GPM (with a special choice of parameters, see Section 3.5.3) give similar results. Taking into account their inherent differences, we summarize that what matters is not a particular 1-particle model, but the degree of irregularity in the 1-particle model chosen.

3.5.3 Gaussian random particles

Let us briefly consider the Gaussian particle model now (Muinonen et al., 1996; Nousiainen and Muinonen, 1999). The surface of the Gaussian random particle is described by the following equation in the spherical coordinate system:

$$r(\vartheta, \phi) = \frac{C}{\sqrt{1+\sigma^2}} \exp\left\{ \sum_{l=0}^{\infty} \sum_{m=0}^{l} [A_{lm} \cos m\phi + B_{lm} \sin m\phi] P_l^m(\cos \vartheta) \right\},$$

where $C = af(\vartheta)$, A_{lm} and B_{lm} are independent Gaussian random variables with variances

$$\beta_{lm}^2 = (2 - \delta_{m0}) \frac{(l-m)!}{(l+m)!} c_l \ln(1 + \sigma^2)$$

and zero means. Here δ_{m0} is the delta function and

$$c_l = (2l+1) \exp(-\varsigma) i_l(\varsigma),$$

where $i_l(\zeta)$ is the modified spherical Bessel function and

$$\zeta = \frac{1}{4\sin^2(\gamma/2)}.$$

We see that this model has three free parameters. These are the average radius a, the coefficient of variance σ, and so-called correlation angle γ. One should also specify the function $f(\vartheta)$. The simplest case is given by $f(\vartheta) = 1$. Nousiainen and Muinonen (1999) and Nousiainen (2000) have used the function

$$f(\vartheta) = 1 + \sum_{n=1}^{10} a_n P_n(\cos\vartheta)$$

with a_n, depending on the size of particles (Nousiainen, 2000). This allowed consideration of the phase matrices of oscillating raindrops. For instance, Nousiainen gives the following values of the coefficients for water droplets with the diameter 1 mm: $a_0 = -0.0028$, $a_1 = 0.003$, $a_2 = -0.0083$, $a_3 = 0.0022$, $a_4 = -0.0003$, $a_5 = -0.0002$, and $a_6 = -0.0001$. The other coefficients are equal to zero.

The results of calculations of the phase function and the normalized matrix at $n = 1.31$, using this model is given in Figure 3.29(a) and (b) together with the results for the FPM. The following parameters have been used in the calculations: $f = 1$, $\sigma = 0.2$, and $\gamma = 5°$. We see that both models give similar results. The results of experimental measurements of the cirrus cloud phase function (von Hoyningen-Huene and Posse, 1997) are also given in Figure 3.29(a). The difference between the theory and measurements at large angles is most probably due to the presence of spherical particles in the scattering volume.

The FPM is more suitable for the solution of the scattering problem by irregular particles. From the general point of view, it is clear that a Gaussian random particle model at a certain choice of parameters (and also other models with extreme randomness) and a FPM should produce similar results. This is confirmed by the comparison of matrices obtained for the FPM and the Gaussian particle model (see Figure 3.29). Note that for a fractal model there are no free parameters because for non-absorbing ice particles the scattering does not depend on the size if the particle shape and refractive index are fixed. Therefore, the model of fractal particles, having fewer free parameters, is preferable for dealing with extremely irregular shapes such as those that occurred in ice clouds.

It is known that strongly absorbing, randomly oriented, non-spherical particles have phase functions close to that of strongly absorbing spheres (independent of their actual shape). This is because in this case the phase matrix outside the diffraction peak is determined only by a reflection process. It seems that there is the same limit for non-absorbing particles of complex shapes. Here, particles of different but complex shapes exhibit common light scattering and polarization features. In this case, however, not only reflected but also refracted light components should be distributed at random.

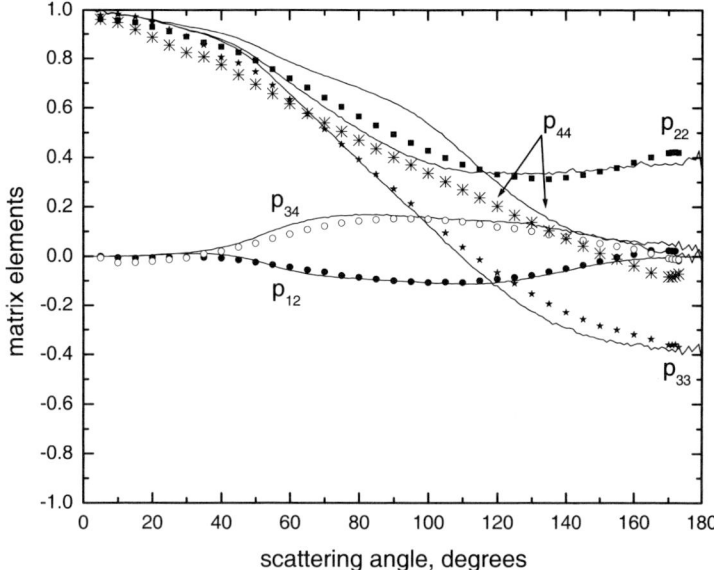

Figure 3.26. Normalized matrix elements, obtained from measurements (symbols from Volten et al., 2001) and calculations at the complex refractive index $m = 1.4 - i0.005$ with $D = 100\,\mu m$ and $\lambda = 0.5\,\mu m$.

vector radiative transfer studies in the ocean. The advantage of this model is due to the fact that it is defined by just one number (n). The correspondence of the model to experimental data can be improved, taking into account other parameters, which influence light scattering in the ocean.

Dust aerosol

We conclude the investigation of the validity of the FPM as compared to experiments by studying the case of yet another medium with highly irregular particles, namely dust aerosols. The main component of the dust aerosol over deserts is well represented by quartz aerosol particles. Measurements of the phase matrices of quartz particles in the visible were reported by Volten et al. (2001). The comparison is given in Figure 3.26, where we assumed that the imaginary part of aerosol particles is equal to 0.005 and $n = 1.4$. The correspondence between experiment and theory is quite good for all matrix elements (except p_{44} in the forward hemisphere). Note that the value of p_{34} is positive as compared to the case of non-absorbing aerosols (compare Figures 3.24 and 3.26). Thus the sign of the element p_{34} (if properly defined) can serve as an indication of light absorption inside scattering particles.

The poor correspondence between experiment and theory for the element p_{44} can be explained by the difference between the actual quartz refraction and absorption coefficients and the assumed values.

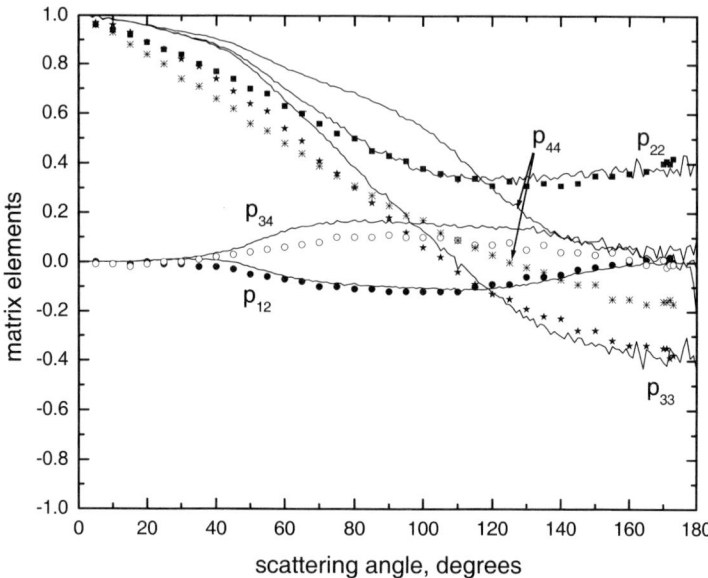

Figure 3.27. The same as in Figure 3.26, but for the averaged normalized phase matrix of aerosol particles (from Volten et al., 2001). Theoretical curves differ from those in Figure 3.26 due to the different statistics, used in a Monte Carlo calculation. We used 1000 orientations of the fractal particle for this figure as opposite to 10 000 orientations of the particle in Figure 3.26. The number of rays used was 1000 in both cases.

Phase functions obtained from theory and experiment (Volten et al., 2001) for quartz particles are compared in Figure 3.26. We see excellent agreement of the FPM to the experimental data.

Volten (2001) and Volten et al. (2001) also gave the average aerosol phase matrix, taking into account measurements of phase matrices by various samples, including Feldspar, Red Clay, Lokon Ash, Loess, Pinatubo Ash, Quartz, and Sahara dust at wavelengths 441.6 and 632.8 nm, which totals 14 measurements. This matrix is given in Figure 3.27 together with results obtained from the FPM, which are identical to that given in Figure 3.27. The agreement of this average matrix model, advised for use in aerosol remote sensing problems (Volten et al., 2001), with the FPM is rather good (with the exception of the element p_{44}).

Phase functions obtained from the theory and averaged experimental data (Volten et al., 2001) for the aerosol are compared in Figure 3.28. We see excellent agreement of the FPM to the experimental data.

This suggests that the FPM may be useful in the aerosol optics studies. Note that Volten et al. (2001) compared their measurements with model computations assuming a stochastic particle model, which differs from the FPM considered above. In particular, the Gaussian random particle model was used (Muinonen et al., 1996). Volten et al. (2001) found that the Gaussian particle model (GPM) describes their data quite well (although theoretical calculations give higher values of

Sec. 3.5] Irregularly shaped particles 125

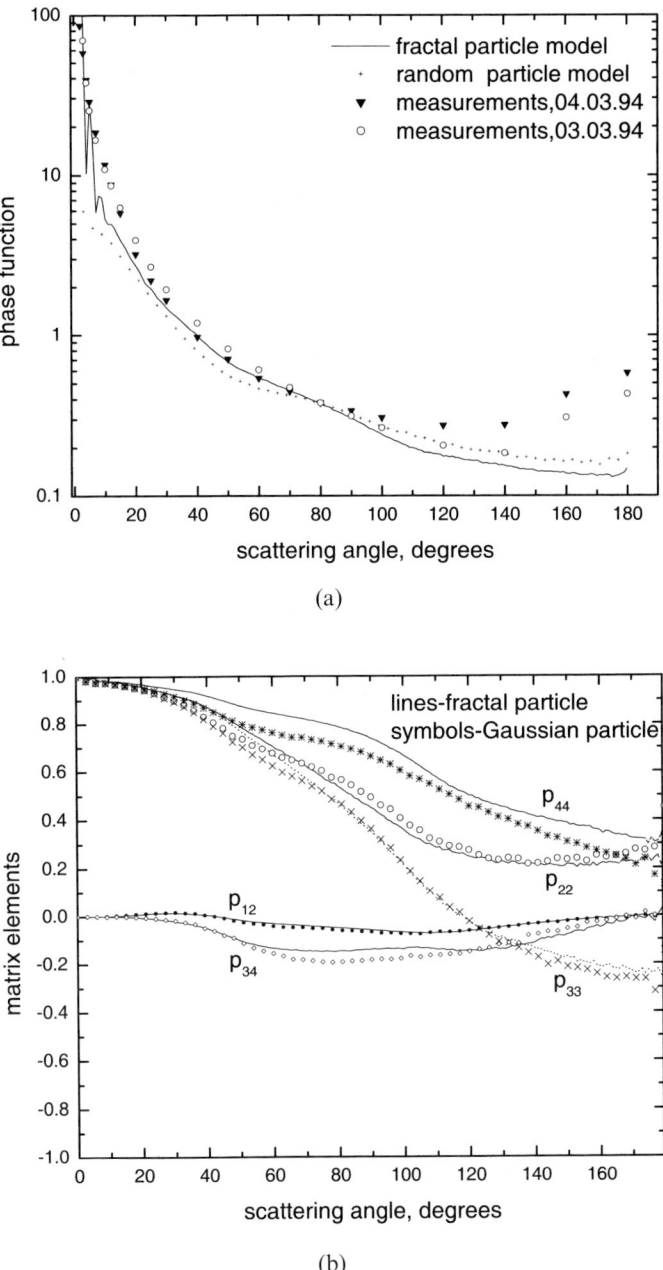

Figure 3.29. (a) The comparison of fractal and Gaussian particle models for the calculation of the phase function at $n = 1.31$. The results of measurements (von Hoyningen-Huene and Posse, 1997) are given by circles and triangles. The parameters of the Gaussian particle are $\sigma = 0.2$ and $\gamma = 5°$ (data for Gaussian model Courtesy of T. Nousiainen, pers. commun.); (b) the same as in (a) except for the normalized phase matrix elements.

3.6 INHOMOGENEOUS PARTICLES

3.6.1 General equations

All the different particles considered above have one similarity, namely, all of them were composed of one substance. In practice, however, particles can be heterogeneous. They often have continuous or discontinuous films on their surfaces, inclusions of foreign particles, a continuous radial variation of the refractive index, or can be composed of several layers. There are special books focused on this topic (e.g., Babenko et al., 2003). On the other hand, I do not like to avoid this subject altogether. So I briefly consider here the most simple type of heterogeneous particles, that is the case of two-layered concentric spheres. The Mie-type solution of this problem was obtained by Aden and Kerker (1951) and Shifrin (1952).

Due to the spherical symmetry of light scattering by a two-layered sphere the expressions (Equations 3.56 and 3.57) for uniform spheres remain valid, but amplitude coefficients a_l and b_l should be changed accordingly. This is also the case for arbitrary numbers of concentric layers, which allows particles with a continuously varying index of refraction to be considered by using a large number of extremely thin layers.

In particular, we have for two-layered spheres (Bohren and Huffman, 1983):

$$a_l = \frac{\psi_l(y)[\psi_l'(m_2 y) - A_l \chi_l'(m_2 y)] - m_2 \psi_l'(y)[\psi_l(m_2 y) - A_l \chi_l(m_2 y)]}{\xi_l(y)[\psi_l'(m_2 y) - A_l \chi_l'(m_2 y)] - m_2 \xi_l'(y)[\psi_l(m_2 y) - A_l \chi_l(m_2 y)]},$$

$$b_l = \frac{m_2 \psi_l(y)[\psi_l'(m_2 y) - B_l \chi_l'(m_2 y)] - \psi_l'(y)[\psi_l(m_2 y) - B_l \chi_l(m_2 y)]}{m_2 \xi_l(y)[\psi_l'(m_2 y) - B_l \chi_l'(m_2 y)] - \xi_l'(y)[\psi_l(m_2 y) - B_l \chi_l(m_2 y)]},$$

where

$$A_l = \frac{m_2 \psi_l(m_2 x) \psi_l'(m_1 x) - m_1 \psi_l'(m_2 x) \psi_l(m_1 x)}{m_2 \chi_l(m_2 x) \psi_l'(m_1 x) - m_1 \chi_l'(m_2 x) \psi_l(m_1 x)},$$

$$B_l = \frac{m_2 \psi_l(m_1 x) \psi_l'(m_2 x) - m_1 \psi_l(m_2 x) \psi_l'(m_1 x)}{m_2 \chi_l'(m_2 x) \psi_l(m_1 x) - m_1 \psi_l'(m_1 x) \chi_l(m_2 x)}$$

and m_1, m_2 are relative refractive indices of the core and shell, respectively, $x = ka$, $y = kb$, $k = 2\pi/\lambda$, a is the radius of the core and b is the radius of the particle.

We see that no new special functions arise in this problem. So the computer code can be easily constructed by modifying the standard Mie code.

3.6.2 Rayleigh approximation

In the Rayleigh approximation for a two-layered spherical particle (Kokhanovsky, 2001):

$$S_1(\theta) = \tfrac{3}{2} a_1 \cos\theta,$$

$$S_2(\theta) = \tfrac{3}{2} a_1,$$

where

$$a_1 = \frac{2i}{3} y^3 F(m), \quad F(m) = \frac{(m_2^2 - 1)(m_1^2 + 2m_2^2) + q^3(2m_2^2 + 1)(m_1^2 - m_2^2)}{(m_2^2 + 2)(m_1^2 + 2m_2^2) + q^3(2m_2^2 - 2)(m_1^2 - m_2^2)}, \quad q = \frac{a}{b}.$$

Formulae for S_1 and S_2 for two-layered spheres differs from those for uniform spheres only by the multiplier a_1 in the framework of the Rayleigh approximation. Therefore, the phase function and normalized matrix elements of uniform and two-layered spheres coincide, assuming that $a_1 \neq 0$. This conclusion is easily generalized with an arbitrary mumber of layers.

The values of $S_1(\theta)$ and $S_2(\theta)$ of Rayleigh particles can be expressed through the polarizability α (Van de Hulst, 1981):

$$S_1(\theta) = ik^3 \alpha \cos\theta, \qquad S_2(\theta) = ik^3 \alpha.$$

Thus, one obtains for the polarizability of a two-layered sphere (Van de Hulst, 1981):

$$\alpha = b^3 F(m).$$

The value of α can be used to calculate the absorption and scattering cross sections of two-layered Rayleigh particles

$$C_{abs} = -4\pi k \operatorname{Im}(\alpha),$$

$$C_{sca} = \frac{8}{3} \pi k^4 |\alpha|^2.$$

It should be noted that at $F(m) = 0$: $\alpha = 0$, $C_{abs} = 0$, and $C_{sca} = 0$. Thus, the particle does not scatter or absorb radiation. It does not act as an optical inhomogeneity in this case. This property can be used in technological applications. If the denominator of $F(m)$ is close to zero, the polarizability α of a two-layered particle is large and there is a spike in the spectral dependencies of the cross sections $C_{abs}(\lambda)$ and $C_{sca}(\lambda)$.

The case of a small coated ellipsoid compared to the wavelength was studied by Bohren and Huffman (1983). It was shown that polarizabilities α_j of coated ellipsoids can be calculated from the following equation:

$$\alpha_j = \frac{V\{(\varepsilon_2 - \varepsilon_m)[\varepsilon_2 + (\varepsilon_1 - \varepsilon_2)(L_j^{(1)} - fL_j^{(2)})] + q^3 \varepsilon_2 (\varepsilon_1 - \varepsilon_2)\}}{[\varepsilon_2 + (\varepsilon_1 - \varepsilon_2)(L_j^{(1)} - fL_j^{(2)})][\varepsilon_m + (\varepsilon_2 - \varepsilon_m) L_j^{(2)}] + q^3 L_j^{(2)} \varepsilon_2 (\varepsilon_1 - \varepsilon_2)},$$

where V is the volume of an ellipsoid, $\varepsilon_1, \varepsilon_2$, and, ε_m are dielectric permittivities of the core, shell, and host medium respectively, $q = \sqrt{(a_1 b_1 c_1 / a_2 b_2 c_2)}$, $a_1, b_1, c_1 (a_2, b_2, c_2)$ are semi-axes of the core (shell), and

$$L_1^{(k)} = \frac{a_k b_k c_k}{2} \int_0^\infty \frac{ds}{(a_k^2 + s) f_k(s)},$$

$$L_2^{(k)} = \frac{a_k b_k c_k}{2} \int_0^\infty \frac{ds}{(b_k^2 + s) f_k(s)},$$

$$L_3^{(k)} = \frac{a_k b_k c_k}{2} \int_0^\infty \frac{ds}{(c_k^2 + s) f_k(s)}$$

are geometrical factors. Here $f_k(s) = \sqrt{(a_k^2 + s)(b_k^2 + s)(c_k^2 + s)}$.

It is easy to show that the values of α and $|\alpha|^2$ in the formulae for the amplitude functions, absorption, and scattering cross sections can be found from the following equations in the case of randomly oriented particles:

$$\alpha = \frac{\alpha_1 + \alpha_2 + \alpha_3}{3}, \qquad |\alpha|^2 = \frac{|\alpha_1|^2 + |\alpha_2|^2 + |\alpha_3|^2}{3}.$$

Clearly, it follows: that $\alpha \equiv \alpha_1$, if the applied electric field is parallel to the axis a_1 of an ellipsoid. Multilayered Rayleigh ellipsoids were studied by Farafonov (2000).

3.6.3 Rayleigh–Gans approximation

The amplitude scattering matrix $\hat{S}(\theta)$ of a coated sphere in the framework of the Rayleigh–Gans approximation has the following form:

$$\hat{S}(\theta) = ik^3 \alpha \begin{pmatrix} \cos\theta & 0 \\ 0 & 1 \end{pmatrix} R(\theta),$$

where the value of the polarizability α is presented in the previous section and

$$R(\theta) = R_b(\theta) + \frac{m_1 - m_2}{m_2 - 1} q^3 R_a(\theta).$$

Here m_1 and m_2 are relative refractive indices of the core and mantle, respectively. The refractive index of the host medium is equal to one. Values of $R_a(\theta)$ and $R_b(\theta)$ are defined by the following equations (Kokhanovsky, 2001):

$$R_a(\theta) = 3 \frac{\sin u - u \cos u}{u^3}, \qquad R_b(\theta) = 3 \frac{\sin u' - u' \cos u'}{u'^3},$$

where $u = 2ka \sin(\theta/2)$, and $u' = 2kb \sin(\theta/2)$. The function $R(\theta)$ can be used to find the values of g, C_{sca}, and $\hat{P}(\theta)$ for two-layered spheres.

The normalized phase matrices for uniform and inhomogeneous particles coincide in this approximation. The phase function, however, differs from the uniform sphere model and is given by

$$p(\theta) = \varsigma(1 + \cos^2\theta)\Phi(\theta),$$

where

$$\Phi(\theta) = \left| R_b(\theta) + \left(\frac{m_1 - m_2}{m_2 - 1}\right) q^3 R_a(\theta) \right|^2$$

and ς is the normalization constant.

The absorption cross section in this approximation is represented by the following formula:

$$C_{abs} = \gamma_1 V_1 + \gamma_2 (V - V_1),$$

where $\gamma_i = (4\pi\chi_i/\lambda)$, V_1 and V_2 are volumes of the core and mantle respectively, and χ_1 and χ_2 are imaginary parts of their refractive indices.

The equations presented above can easily be generalized for the case of an N-layered particle with radius b:

$$\Phi(\theta) = \left| R_N(\theta) + \sum_{i=1}^{N-1} \frac{m_i - m_{i+1}}{m_N - 1} \frac{V_i}{V_n} R_i(\theta) \right|^2,$$

$$C_{abs} = \sum_{i=1}^{N} \gamma_i V_i,$$

where V_i and $m_i = n_i - \chi_i$ are the volume and refractive index of the ith layer respectively, and $\gamma_i = (4\pi\chi_i/\lambda)$. The form factors $R_i(\theta)$ and $R_N(\theta)$ are determined by the formulae given above, where u is equal to $2ka_i\sin(\theta/2)$ or $2kb\sin(\theta/2)$, respectively. Here a_i is the inner radius of the ith layer. The equations presented here can also be used for calculations of local optical characteristics of non-uniform spherically symmetrical particles with permanently varying refractive indices $m(r)$ (r is the coordinate along the particle radius). In this case one should divide the particle into N thin layers, where $N \to \infty$.

3.6.4 Polydispersions of coated spheres

The optical properties of two-layered spherical particles with arbitrary radius and refractive index can be studied, using exact electromagnetic theory. The results of numerical calculations for polydispersions of coated spheres are given in Figure 3.30.

Calculations have been performed for the Junge particle size distribution

$$f(a) = Aa^{-\nu},$$

where

$$A = \frac{a_s^\nu a_f^\nu \nu}{a_f^\nu - a_s^\nu}.$$

130 **Local optical characteristics** [Ch. 3

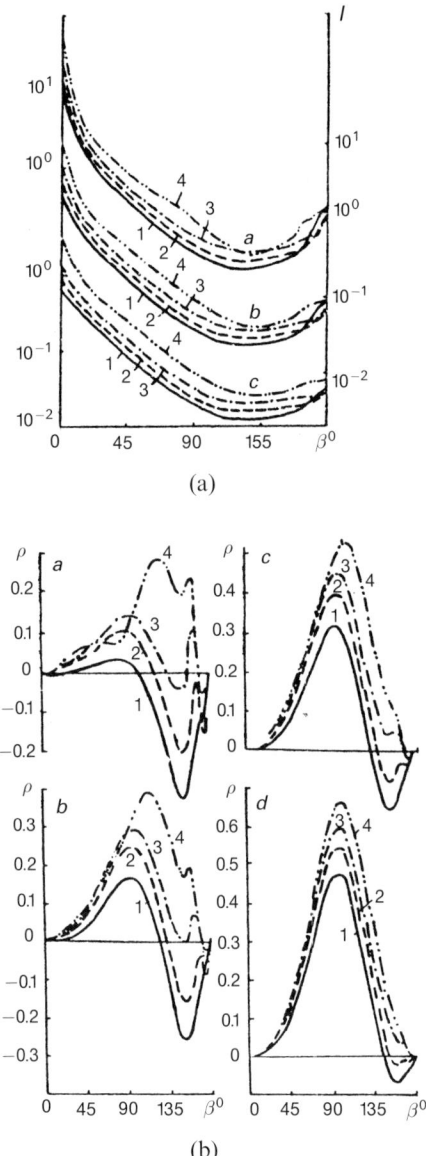

Figure 3.30. (a) The light scattered intensity I for spherical polydispersions of coated spheres with outer to inner diameter ratio equal to 1(1), 1.125(2), 1.27(3), and 1.55(4) at the wavelength 0.6328 μm (Prishivalko, 1976) as function of the scattering angle β. It is assumed that particles radii are distributed according to the Junge law $f(a) = Aa^{-\nu}$ with $\nu = 2.5(a)$, 3.0(b), and 3.5(c). The refractive index of homogeneous spheres is equal to $1.65165 - i0.005$. The refraction index of two-layered particles is modelled as explained in the text (Prishivalko, 1976). The scale to the right (vertical axis) corresponds to case 'b'. Upper scale to the left (vertical axis) corresponds to the case 'a' and 'c'; (b) the same as in (a) but for the degree of linear polarization. The case 'd' corresponds to $\nu = 4$ (Prishivalko, 1976).

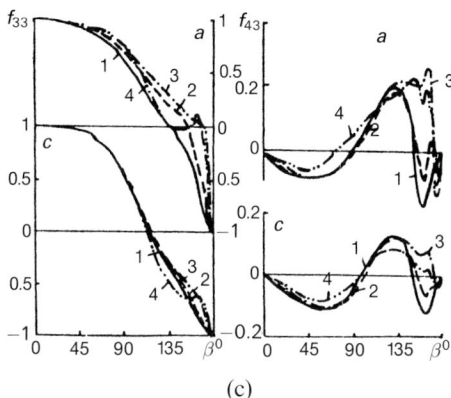

(c)

Figure 3.30 (*cont.*). (c) the same as in (a) but for the normalized phase matrix elements $p_{33} = f_{33}$ and $p_{43} = -p_{34} = f_{43}$. The case of $\nu = 3.0$ (b) is not shown (Prishivalko, 1976).

Here a_s and a_f are the smallest and largest radii of the particles' nuclei. It was assumed that the radii of coated spheres b are proportional to the radii of the nuclei, $b = \varsigma a$.

The values of ς used in the calculations were 1.0 (the case of a uniform sphere), 1.125, 1.27, and 1.55. Other parameters were assumed to have the following values, $\nu = 2.5$ and 5, $\lambda = 0.6328\,\mu\text{m}$ (laser wavelength), $a_s = 0.04\,\mu\text{m}$, and $a_f = 3\,\mu\text{m}$. The refractive index of the nuclei was equal to 1.65–0.005i and the refractive index of the shell $\tilde{m} = \tilde{n} - i\tilde{\chi}$ was assumed to be dependent on the value of ς and equal to 1.65–0.005i at $\varsigma = 1$, 1.487–0.0024i at $\varsigma = 1.27$, and 1.417–0.0013i at $\varsigma = 1.55$. This model describes the decrease of the refractive index of the shell with humidity due to the finite size of a nucleus as was suggested by Hanel (1972). Therefore, for higher humidity we have larger particles with a refractive index approaching that of water (1.33) as b grows and $a = \text{const}$. Note that values of ν of 2.5–4.0 are typical for atmospheric aerosols.

Therefore, the results presented in Figure 3.30 are not just of academic interest, but they model the change of the phase matrix of aerosol particles in a humid atmosphere due to adsorption of water vapor molecules on their surfaces. The cases ς equal to 1.0, 1.125, 1.27, and 1.55 correspond to the relative humidity v equal to 0, 60, 80, and 92% respectively according to the model of Hanel (1972).

The main conclusions are as follows. An increase of humidity also leads to an increase of light scattered intensity. The degree of polarization is positive for most angles. It becomes positive for all scattering angles for high values of humidity. The element p_{33} is larger at $\varsigma = 1$. The value of the scattering angle, where p_{33} is equal to zero, depends on humidity at $\nu = 2.5$. This not the case for $\nu = 3.5$. The behaviour of p_{34} on the scattering angle only weakly depends on the humidity in the forward hemisphere. However, p_{34} changes dramatically with ν in the backward hemisphere. This can be used for remote sensing applications.

3.7 OPTICALLY ACTIVE PARTICLES

3.7.1 General equations

To end this Chapter we briefly discuss the scattering properties of optically active particles. They differ from the particles studied above because they have different refractive indices for left- and right-handed circularly polarized waves. Only the case of spheres will be studied here. However, exact results can also be obtained for other types of optically achive particles, including layered spheres, uniform and layered cylinders, and spheroids (Bohren, 1974, 1975a,b, 1978; Cooray and Ciric, 1993).

Most bio-particles, including red blood cells, membranes, virus nuclei, and mitochondria are characterized by asymmetry of their molecular structure and shape. This asymmetry is one of the main features of living things and has many interesting manifestations. For instance, absorption and extinction by such particles differs with the sense of rotation of incident circularly polarized light. This is not the case for inorganic spheres (e.g., water droplets). The plane of polarization of linearly polarized light rotates during propagation in such asymmetric media. This is due to the different speed of right-handed and left-handed circularly polarized waves in chiral or optically active media.

Electromagnetic scattering by optically active particles differs from the correspondent scattering problem for non-optically active particles due to the necessity to use different material equations. Indeed, for chiral media (Fedorov, 1976):

$$\vec{D} = \varepsilon \vec{E} + \gamma \varepsilon \vec{\nabla} \times \vec{E}, \qquad \vec{B} = \mu \vec{H} + \gamma' \mu \vec{\nabla} \times \vec{H},$$

where ε and μ are the dielectric and magnetic permittivities respectively, $\vec{E}$ and $\vec{H}$ are the electric and magnetic vectors, and $\vec{D}$ and $\vec{B}$ are the dielectric displacement and magnetic induction, respectively. Note that we will consider only isotropic chiral media and assume that values of ε, μ, and $\gamma = \gamma'$ are scalars. Light scattering by anisotropic particles has been studied by Babenko et al. (2003). Parameters γ and γ' are responsible for the chirality of a medium. One can see that it follows at $\gamma = \gamma' = 0$:

$$\vec{D} = \varepsilon \vec{E}, \qquad \vec{B} = \mu \vec{H}.$$

Thus, the general solutions of the scattering problem at $\gamma \neq 0$ and $\gamma' \neq 0$ reduce to the standard and well known solutions at spectral regions where $\gamma = \gamma' = 0$.

The dimensionless phase matrix $\hat{P}$ has the following form for chiral spheres (Bohren and Huffman, 1983):

$$\hat{P} = w \begin{pmatrix} \frac{i_1 + i_2}{2} + i_3 & \frac{i_1 - i_2}{2} & \operatorname{Re}(\mu - \nu) & \operatorname{Im}(\mu + \nu) \\ \frac{i_1 - i_2}{2} & \frac{i_1 + i_2}{2} - i_3 & \operatorname{Re}(\mu + \nu) & \operatorname{Im}(\mu - \nu) \\ -\operatorname{Re}(\mu - \nu) & -\operatorname{Re}(\mu + \nu) & \operatorname{Re}(\delta) - i_3 & \operatorname{Im}(\delta) - i_3 \\ \operatorname{Im}(\mu + \nu) & \operatorname{Im}(\mu - \nu) & i_3 - \operatorname{Im}(\delta) & \operatorname{Re}(\delta) + i_3 \end{pmatrix},$$

where $i_1 = |S_{11}|^2$, $i_2 = |S_{22}|^2$, $i_3 = |S_{12}|^2$, $\mu = S_{11} S_{12}^*$, $\nu = S_{22} S_{12}^*$, $\delta = S_{11} S_{22}^*$, and w is the normalization constant. Explicit expressions for the elements $S_{11}, S_{12} = -S_{21}$, and S_{22} of the amplitude scattering matrix have the following forms (Bohren and Huffman, 1983):

$$S_{11}(\theta) = \sum_{n=1}^{\infty} \frac{2n+1}{n(n+1)} \{g_n \tau_n(\cos\theta) + d_n \pi_n(\cos\theta)\},$$

$$S_{22}(\theta) = \sum_{n=1}^{\infty} \frac{2n+1}{n(n+1)} \{g_n \pi_n(\cos\theta) + d_n \tau_n(\cos\theta)\},$$

$$S_{12}(\theta) = \sum_{n=1}^{\infty} \frac{2n+1}{n(n+1)} c_n (\pi_n + \tau_n),$$

where

$$\pi_n(\cos\theta) = \frac{P_n^{(1)}(\cos\theta)}{\sin\theta}, \quad \tau_n(\cos\theta) = \frac{dP_n^{(1)}(\cos\theta)}{d\theta},$$

$$g_n = \frac{V_n(R)A_n(L) + V_n(L)A_n(R)}{W_n(L)V_n(R) + V_n(L)W_n(R)},$$

$$d_n = \frac{W_n(L)B_n(R) + W_n(R)B_n(L)}{W_n(L)V_n(R) + V_n(L)W_n(R)},$$

$$c_n = i\frac{W_n(R)A_n(L) - W_n(L)A_n(R)}{W_n(L)V_n(R) + V_n(L)W_n(R)},$$

$$W_n(J) = m\psi_n(m_J x)\xi_n'(x) - \xi_n(x)\psi_n'(m_J x),$$

$$V_n(J) = \psi_n(m_J x)\xi_n'(x) - m\xi_n(x)\psi_n'(m_J x),$$

$$A_n(J) = m\psi_n(m_J x)\psi_n'(x) - \psi_n(x)\psi_n'(m_J x),$$

$$B_n(J) = \psi_n(m_J x)\psi_n'(x) - m\psi_n(x)\psi_n'(m_J x).$$

Here

$$\psi_n(x) = \sqrt{\frac{\pi x}{2}} J_{n-(1/2)}(x), \xi_n(x) = \sqrt{\frac{\pi x}{2}} H_{n+(1/2)}(x), J_{n+(1/2)}$$

and $H_{n-(1/2)}$ are the Bessel and Hankel functions and $P_n^{(1)}(\cos\theta)$ is the associated Legendre polynomial. Note that, $\pi_n(0) = \tau_n(0) = n(n+1)/2$, $\pi_n(\pi) = -\tau_n(\pi) = (-1)^{n+1} n(n+1)/2$. Values of J are equal to L or R, $m_L = (N_L/n)$, $m_R = (N_R/n)$, $m = (m_L m_R/\bar{m})$, $\bar{m} = (m_L + m_R)/2$, n is the refractive index of the host medium, N_L and N_R are the refractive indices of the particles for left and right circularly polarized waves, $x = nka$ is the size parameter, $k = 2\pi/\lambda$, and λ is the wavelength. Note that it is supposed that $\gamma = \gamma'$ and the magnetic permittivity of the particles and the host are the same.

It follows that the scattering matrix for chiral spheres is determined by three real numbers i_1, i_2, and i_3 and three complex numbers μ, ν, and δ.

To understand the physical meaning of the different matrix elements one should consider the interaction of a light field, characterized by the specific Stokes vector $\vec{I}_0$, with the medium and characterized by the phase matrix $\hat{P}$ as discussed in Section 3.1.1 (see Equation 3.14). For instance, the Stokes vector $\vec{S}(I,Q,U,V)$ of a singly scattered beam illuminating a turbid chiral layer with horizontal linearly polarized light ($I_0 = Q_0, U_0 = V_0 = 0$) is:

$$I = (i_1 + i_3)I_0, \quad Q = (i_1 - i_3)I_0, \quad U = -2\operatorname{Re}(\mu)I_0, \quad V = 2\operatorname{Im}(\mu)I_0.$$

For isotropic spheres, $i_3 = \mu = 0$, and the state of polarization does not change due to the scattering process. This is not the case for chiral spheres.

It follows that the degree of polarization

$$q = \frac{\sqrt{Q^2 + U^2 + V^2}}{I}$$

of singly scattered light is equal to one as in the case of non-chiral scatterers. However, the state of polarization is not linear, but it is now elliptical ($V \neq 0$). Note that such a type of polarization is common (e.g., in the ocean, which can be considered as a gigantic bio-colloid, composed of both intrinsically symmetric and asymmetric particles).

For the parameters of the polarization ellipse of singly scattered light

$$\psi = \tfrac{1}{2}\arctan\frac{U}{Q}$$

and

$$\varphi = \tfrac{1}{2}\arcsin\frac{V}{I}$$

it follows approximately that:

$$\psi = \frac{\operatorname{Re}(\mu)}{i_1 - i_3}, \quad \varphi = \frac{\operatorname{Im}(\mu)}{i_1 + i_3},$$

where we accounted for the fact that $\mu \ll 1$ due to the small values of the chiral parameter γ. The value of ψ is the azimuth and φ is the ellipticity of the polarization ellipse. We see that $\psi \sim \operatorname{Re}(\mu) = (P_{23} + P_{13})/2$ and $\varphi \sim \operatorname{Im}(\mu) = (P_{14} + P_{24})/2$. This gives physical meaning to elements P_{12}, P_{23}, P_{14}, and P_{24}.

Let us now consider the illumination of a layer by vertical linearly polarized light ($I_0 = -Q_0, U_0 = V_0 = 0$). In this case:

$$I = (i_2 + i_3)I_0, \quad Q = (i_2 - i_3)I_0, \quad U = -2\operatorname{Re}(\nu)I_0, \quad V = 2\operatorname{Im}(\nu)I_0$$

and, if $\nu \ll 1$, we have

$$\psi = \frac{\operatorname{Re}(\nu)}{i_2 - i_3}, \quad \varphi = \frac{\operatorname{Im}(\nu)}{i_2 + i_3}.$$

Note that $\operatorname{Re}(\nu) = (P_{23} + P_{13})/2$ and $\operatorname{Im}(\nu) = (P_{14} + P_{24})/2$. The real and complex parts of ν are responsible for the rotation of the polarization plane and the ellipticity of the scattered radiation, respectively.

Let us now consider the normalized matrix $\hat{p} = \hat{P}/P_{11}$ or

$$\hat{p} = \begin{pmatrix} 1 & b_1 & b_3 & b_5 \\ b_1 & a_2 & b_4 & b_6 \\ -b_3 & -b_4 & a_3 & b_2 \\ b_5 & b_6 & b_2 & a_4 \end{pmatrix}.$$

This matrix can be represented as a sum of two matrices:

$$\hat{p} = \hat{p}' + \hat{p}'',$$

where

$$\hat{p}' = \begin{pmatrix} 1 & b_1 & 0 & 0 \\ b_1 & a_2 & 0 & 0 \\ 0 & 0 & a_3 & b_2 \\ 0 & 0 & b_2 & a_4 \end{pmatrix}$$

and

$$\hat{p}'' = \begin{pmatrix} 0 & 0 & b_3 & b_5 \\ 0 & 0 & b_4 & b_6 \\ -b_3 & -b_4 & 0 & 0 \\ b_5 & b_6 & 0 & 0 \end{pmatrix}.$$

The effects of the chirality on the elements of the matrix $\hat{p}'$ are small and can be neglected in the framework of the first approximation. Thus, it follows that the elements $a_2 \approx 1$ and $a_3 \approx a_4$, and b_1 and b_2 have values that are very close to those that can be obtained from the usual Mie theory with the average refractive index $m = (m_L + m_R)/2$. This is due to the fact that the refractive indices for left-handed m_L and right-handed m_R circularly polarized waves do not differ very much. The same is true for the phase function.

Note that b_2 is equal to zero for non-chiral Rayleigh spheres. However, for chiral particles, $b_2 = -2i_3/(i_1 + i_2 + 2i_3)$, where we have accounted for the fact that $\mathrm{Im}(\delta) = 0$ in the case of Rayleigh scatterers.

The elements b_3, b_4, b_5, and b_6 are equal to zero for non-chiral spheres. For chiral particles this is the case only at $\theta = \pi$, when the element of the amplitude scattering matrix $S_{12} = 0$. The values of b_3 and b_6 are also equal to zero at $\theta = 0$, because $S_{11}(0) = S_{22}(0)$ and $\mu = \nu$ in this case. This is not a case for elements b_4 and b_5.

It follows that $\hat{p}''(\pi) = 0$ and $\hat{p}(\pi) = \hat{p}'(\pi)$. Thus, the influence of the chirality on the matrix

$$\hat{p}(\pi) = \begin{pmatrix} 1 & 0 & 0 & 0 \\ 0 & a_2 & 0 & 0 \\ 0 & 0 & a_3 & 0 \\ 0 & 0 & 0 & a_4 \end{pmatrix},$$

in the backward direction can be neglected. In the forward direction:

$$\hat{p}(0) = \begin{pmatrix} 1 & 0 & 0 & b_5 \\ 0 & a_2 & b_4 & 0 \\ 0 & -b_4 & a_3 & 0 \\ b_5 & 0 & 0 & a_4 \end{pmatrix},$$

where the value of b_5 is due to the circular dichroism and the value of b_4 is due to the optical rotation in the forward direction. Thus, the chirality can be ignored in the backward but not in the forward direction.

We give the phase function $p(\theta)$ and normalized phase matrix elements in Figure 3.31 for the case of the poly-L-glutamic acid (PGA) at $\lambda = 0.19\,\mu\mathrm{m}$. It is assumed that the PSD is given by Equation 1.5 at $\mu = 6$. We also use, $n_L = 1.723\,438$, $\chi_L = 0.140\,046$, $n_R = 1.723\,161$, $\chi_R = 0.140\,046$ (Bohren, 1975a). It follows that the phase function becomes more extended in the forward direction with the size of particles. The Rayleigh approximation (see Section 3.7.3) can be used starting from the effective radius equal to $0.01\,\mu\mathrm{m}$ and lower. The values of

$$\psi = \frac{P_{13} + P_{23}}{2(P_{12} + P_{22})},$$

$$\varphi = \frac{P_{14} + P_{24}}{2(P_{12} + P_{11})}$$

give the optical rotatory dispersion (ORD) and circular dichroism (CD) values of scattered light for a horizontally polarized light beam incident on a turbid layer (see Figures 3.31(i), and (j)) at the wavelength $0.19\,\mu\mathrm{m}$. Generally, ORD and CD spectra depend on the scattering angle and particle size. Both values ψ and φ decrease with angle. It is interesting to note that for Rayleigh chiral spheres: $\varphi(\pi/2) = \psi(\pi/2) = 0$.

In principle, the dependencies $\varphi(\theta, \lambda)$ and $\psi(\theta, \lambda)$ can be used to obtain the intrinsic ORD and CD spectra of substances in a dispersed state.

3.7.2 The extinction matrix

For the extinction matrix of chiral spheres (Kokhanovsky, 2001):

$$\hat{\sigma}_{ext} = \varepsilon \begin{pmatrix} 1 & 0 & 0 & \beta \\ 0 & 1 & \alpha & 0 \\ 0 & -\alpha & 1 & 0 \\ -\beta & 0 & 0 & 1 \end{pmatrix}$$

Sec. 3.7] **Optically active particles** 137

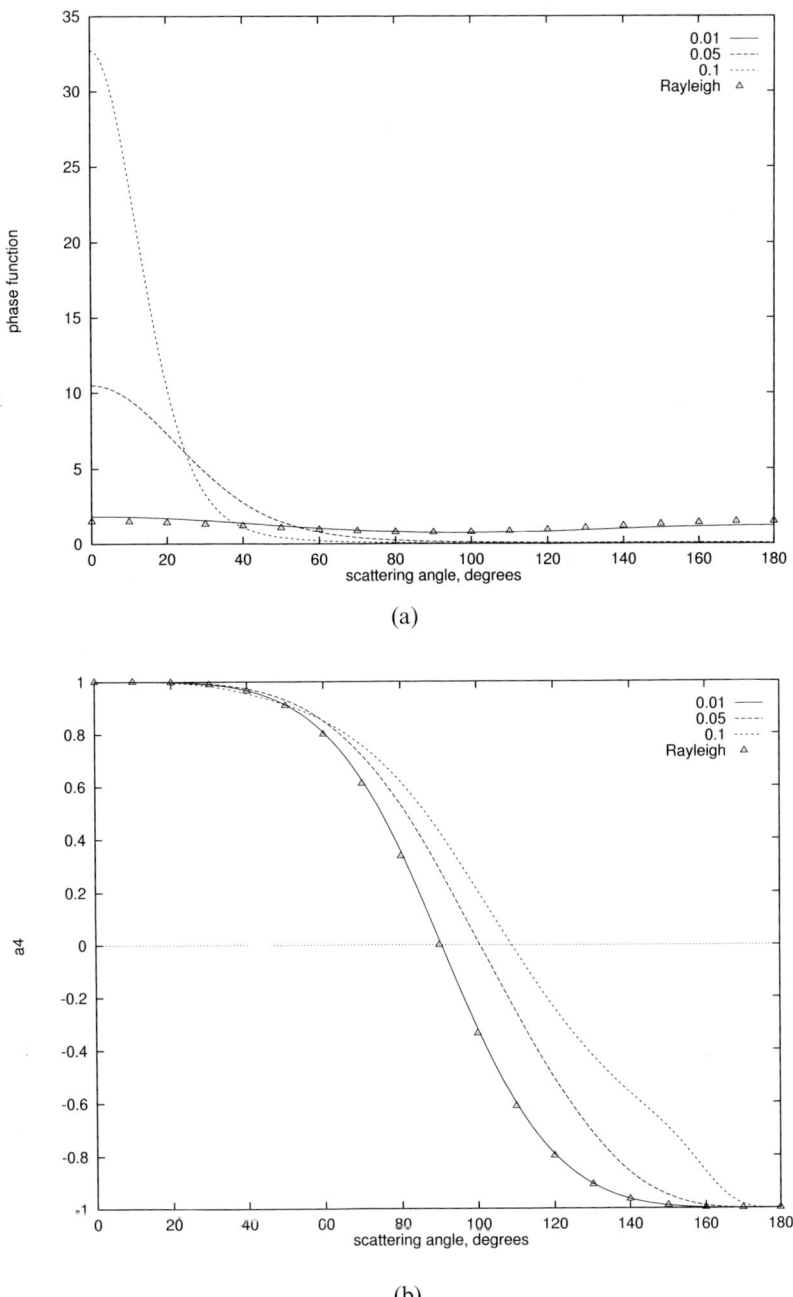

Figure 3.31. The angular dependence of the phase function (a), the normalized phase matrix elements (b–h), and the optical rotatory dispersion (ORD) (i) and circular dichroism (CD) (j) spectra of PGA spheres for the effective radius of particles equal to 0.01, 0.05, and 0.1 μm. The wavelength is equal to 0.19 μm. Results for chiral Rayleigh spheres are also given.

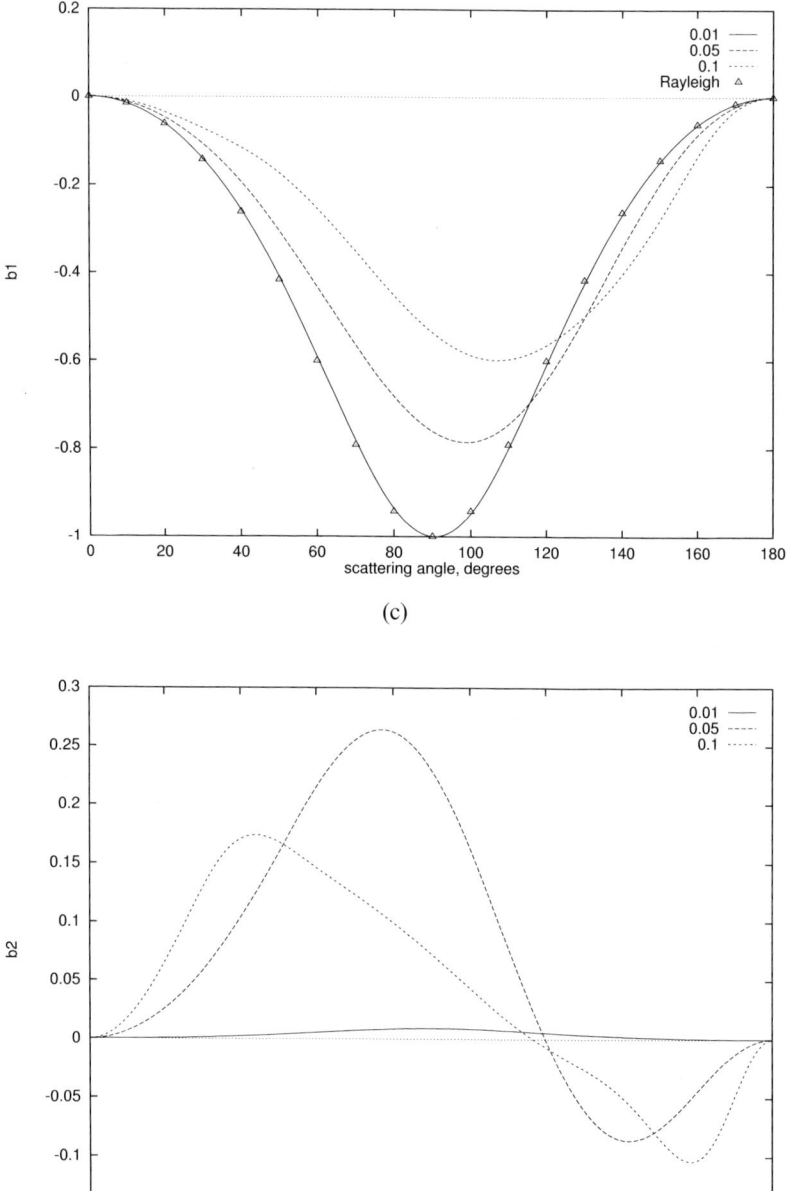

Figure 3.31 (*cont.*)

Sec. 3.7] **Optically active particles** 139

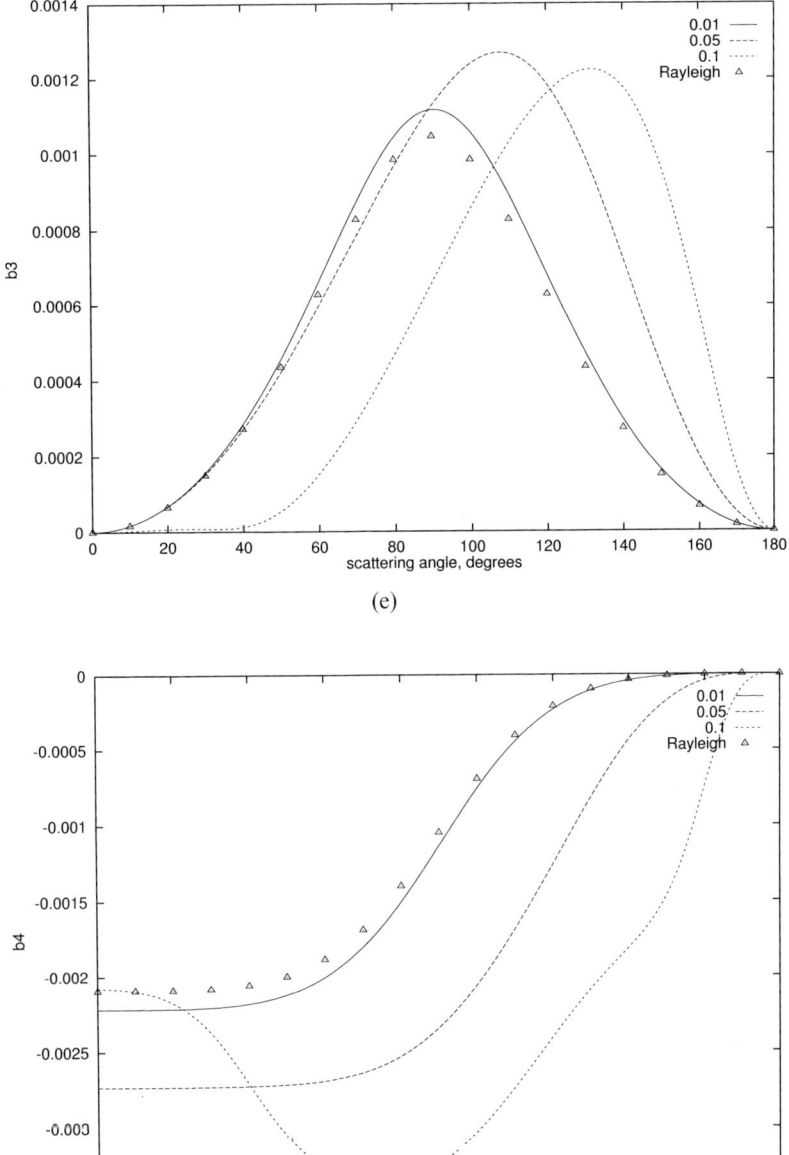

(e)

(f)

Figure 3.31 (*cont.*)

140 Local optical characteristics [Ch. 3

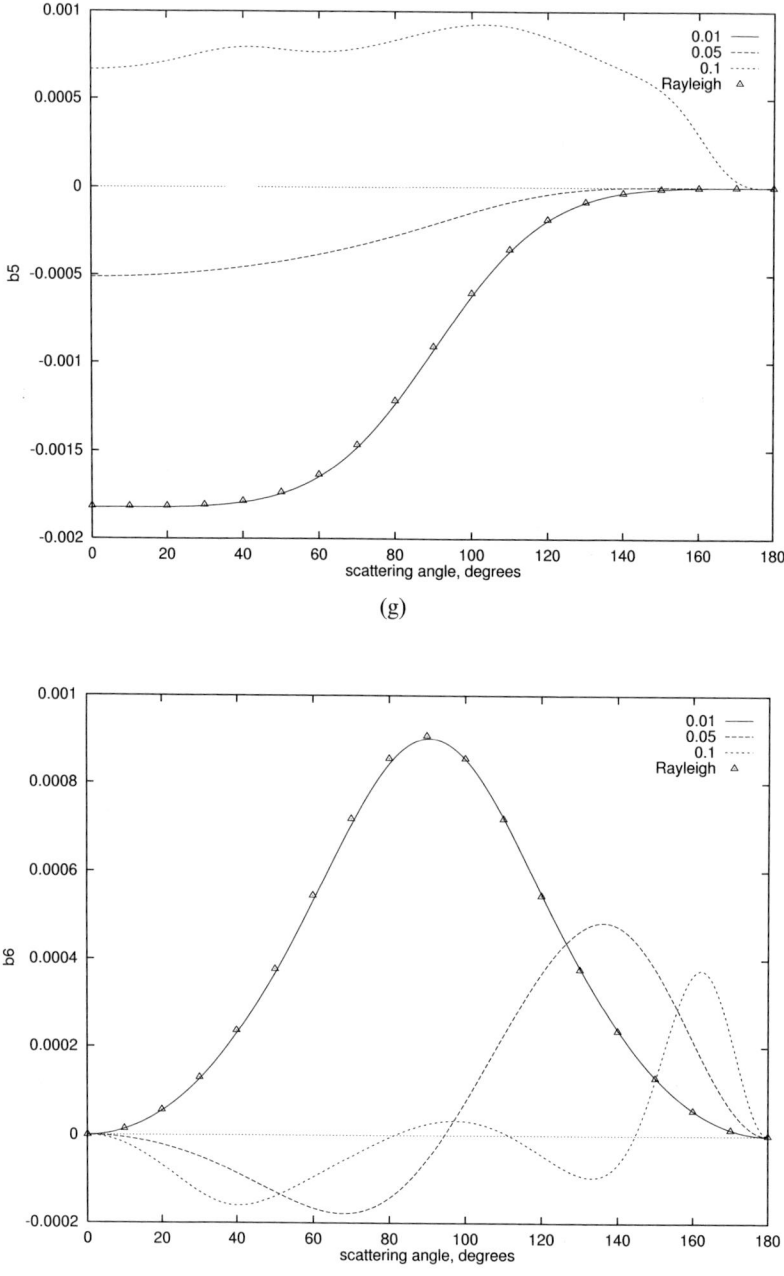

(g)

(h)

Figure 3.31 (*cont.*)

Sec. 3.7] **Optically active particles** 141

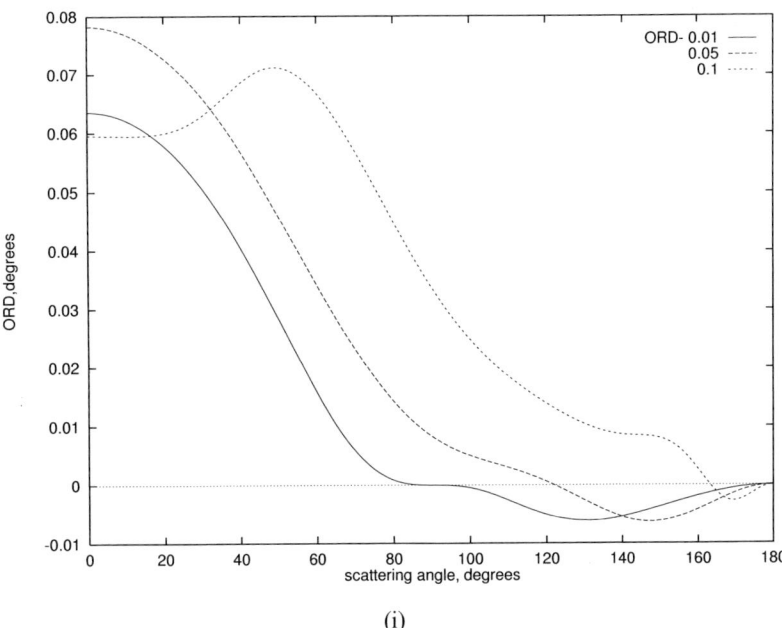

(i)

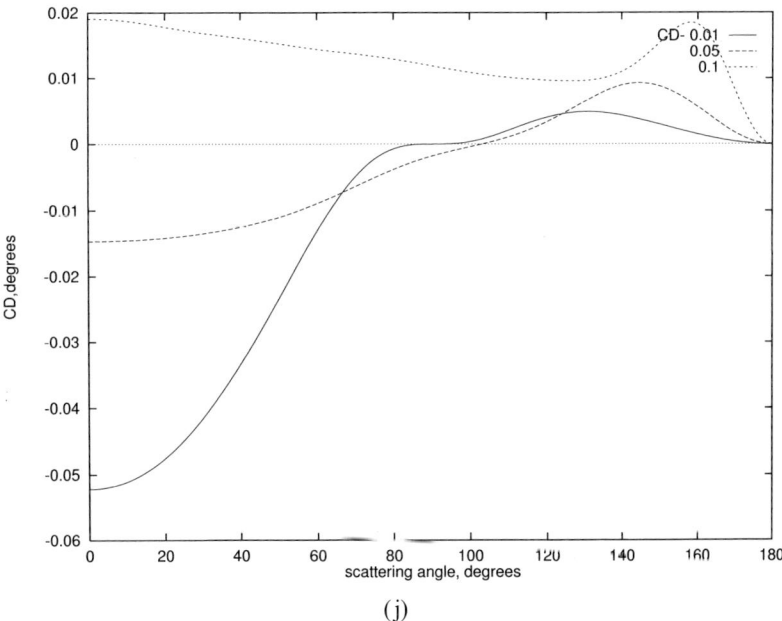

(j)

Figure 3.31 (*cont.*)

where

$$\varepsilon = \frac{4\pi N}{k^2} \operatorname{Re}[S_{11}(0)],$$

$$\alpha = -\frac{\operatorname{Im}[S_{12}(0)]}{\operatorname{Re}[S_{11}(0)]},$$

$$\beta = \frac{\operatorname{Re}[S_{12}(0)]}{\operatorname{Re}[S_{11}(0)]}.$$

Note, that values of α and β are small and decrease with the size of the scatterers.

One can show (Kokhanovsky, 1999) that the rotation of the plane of polarization of a linearly polarized incident beam is given:

$$\psi = \frac{\beta s}{2},$$

where $s = \varepsilon z/\mu_0$, z is the geometrical depth, and μ_0 is the cosine of the incidence angle. It is possible to obtain for the ellipticity angle (Kokhanovsky, 1999):

$$\varphi = -\frac{\alpha s}{2}.$$

To calculate the spectra $\psi(\lambda), \varphi(\lambda)$ we need to know refractive indices of the optically active particles. A convenient model was proposed by Bohren (1975a). He gives the following equations for the refractive indices $m_L = n_L - i\chi_L$ and $m_R = n_R - i\chi_R$:

$$n_L = \bar{n} + \frac{\Delta n}{2}, \qquad \chi_L = \bar{\chi} + \frac{\Delta \chi}{2},$$

$$n_R = \bar{n} - \frac{\Delta n}{2}, \qquad \chi_R = \bar{\chi} - \frac{\Delta \chi}{2},$$

$$\Delta n = \sum_{j=1}^{M} \frac{D_j \lambda (\lambda^2 - \lambda_j^2)}{(\lambda^2 - \lambda_j^2)^2 + \Gamma_j^2 \lambda^2},$$

$$\Delta \chi = \sum_{j=1}^{M} \frac{D_j \Gamma_j \lambda^2}{(\lambda^2 - \lambda_j^2)^2 + \Gamma_j^2 \lambda^2},$$

$$\bar{n} = \operatorname{Re}\left(\sqrt{\varepsilon' + i\varepsilon''}\right),$$

$$\bar{\chi} = \operatorname{Im}\left(\sqrt{\varepsilon' + i\varepsilon''}\right),$$

$$\varepsilon' = \varepsilon_0 + \sum_{j=1}^{M} \frac{F_j \lambda^2 (\lambda^2 - \lambda_j^2)}{(\lambda^2 - \lambda_j^2)^2 + G_j^2 \lambda^2},$$

$$\varepsilon'' = \varepsilon_0 + \sum_{j=1}^{M} \frac{F_j G_j \lambda^3}{(\lambda^2 - \lambda_j^2)^2 + G_j^2 \lambda^2},$$

where values of $F_j, G_j, \lambda_j, D_j, \Gamma_j$, and M depend on the substance in question. We will use the values of the parameters F_j, G_j, D_j, and Γ_j proposed by Bohren

Table 3.9. Parameters of the refractive index parameterization for the case of PGA (Bohren, 1975a).

λ_j (nm)	D_j (nm)	F_j (nm)	Γ_j (nm)	G_j (nm)
190	0.016	0.051	14	20
208	−0.007	0	15	0
224	−0.009	0	15	0

(1975a), assuming that $M = 3$ (see Table 3.9). Then the spectra obtained are close to the experimentally measured spectra for the PGA.

The spectral dependencies of $\bar{n}(\lambda), \bar{\chi}(\lambda), \Delta n(\lambda),$ and $\Delta\chi(\lambda)$ for the case of PGA are presented in Figure 3.32 (Bohren, 1975a). The results of calculations for Rayleigh chiral particles are found in Figure 3.33.

If follows from Figure 3.33(a) that even very small particles ($a_{ef} = 0.005\,\mu\text{m}$) distort PGA spectra in water. For larger particles (see Figures 3.33(b) and (c)) distortions of spectra are more prominent. The maxima move to larger wavelengths and the values of ORD and CD at the maxima reduce. It should be pointed out that a further increase of size can lead to drastic changes in the spectra. The calculated

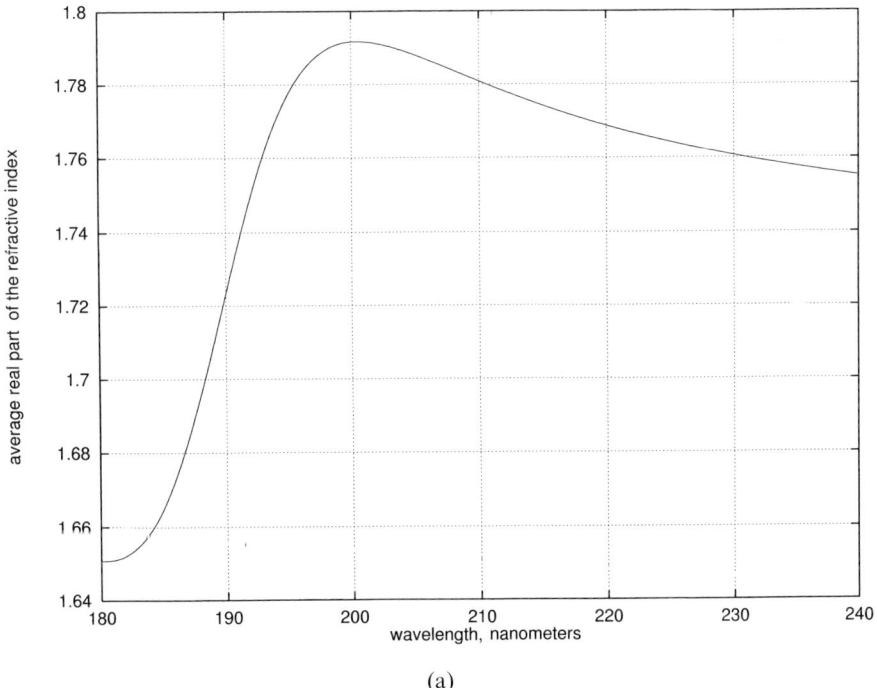

(a)

Figure 3.32. (a) The spectral dependence of the average real part of the refractive index of the PGA.

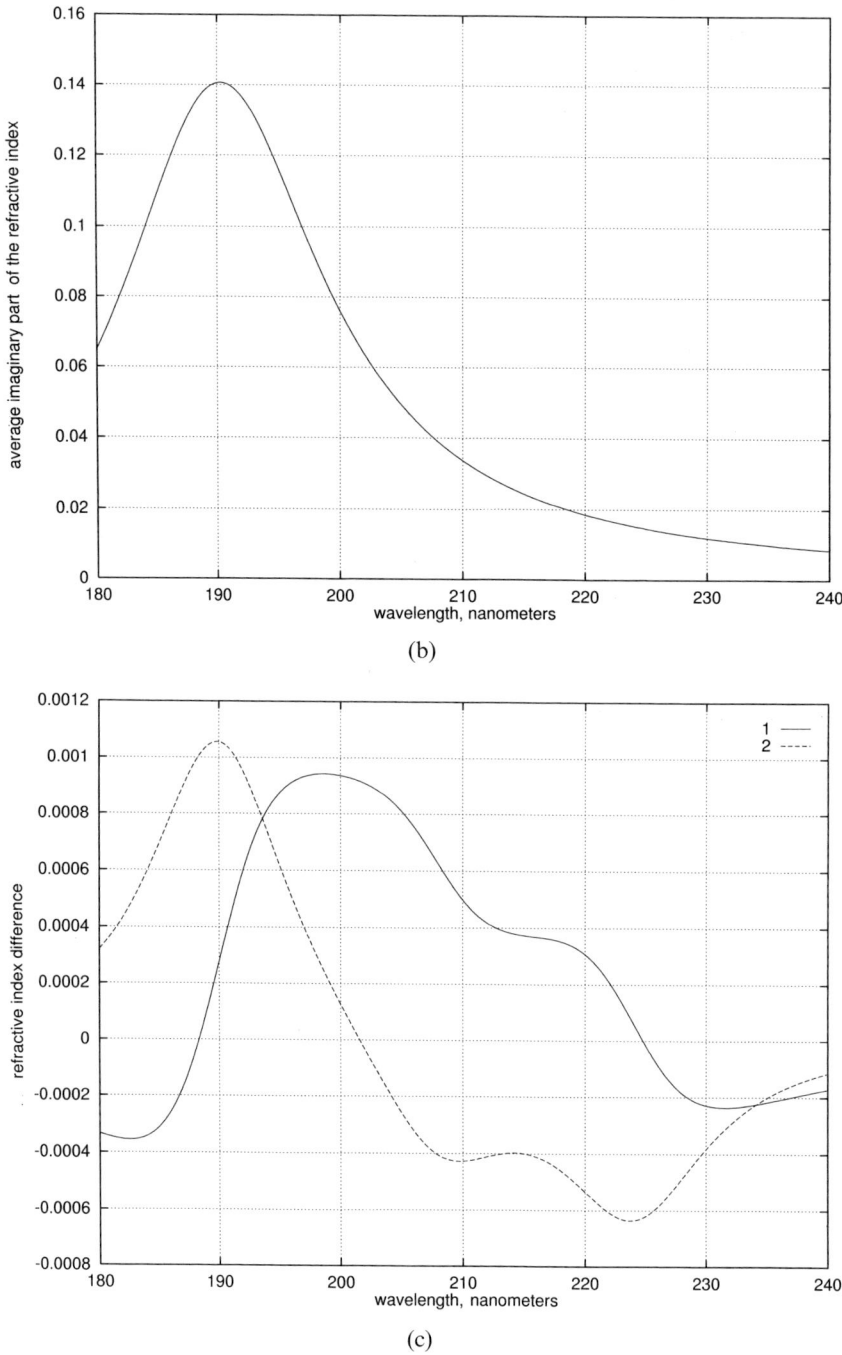

Figure 3.32 (*cont.*). (b) The spectral dependence of the average imaginary part of the refractive index of the PGA; (c) the spectral dependence of Δn (1) and $\Delta \chi$ (2).

Sec. 3.7] Optically active particles 145

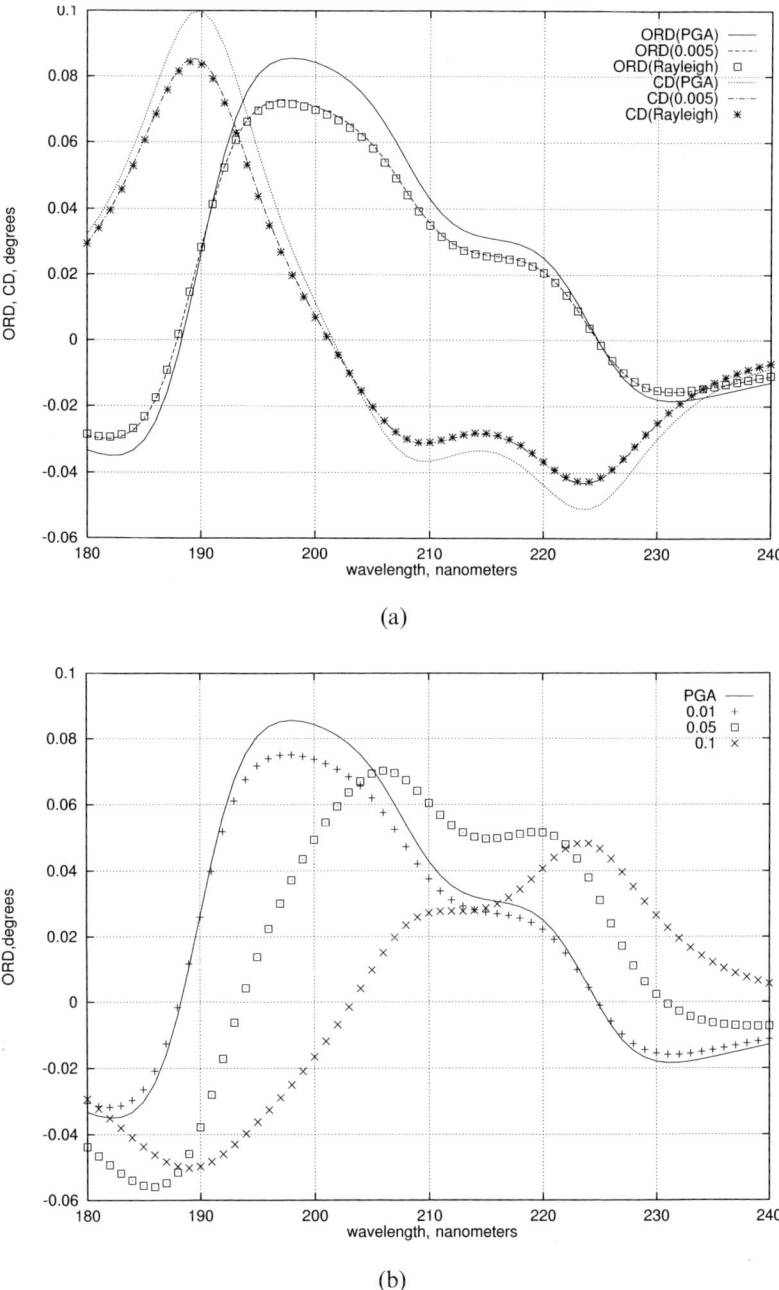

Figure 3.33. (a) The ORD and CD spectra (in transmitted light) of the PGA in solution and particulate media made of the PGA spheres, having the effective radius equal to 0.005 μm and that of a Rayleigh limit; (b) the ORD spectrum (in transmitted light) for the PGA in solution and PGA spheres, having the effective radii equal to 0.01, 0.05, and 0.1 μm.

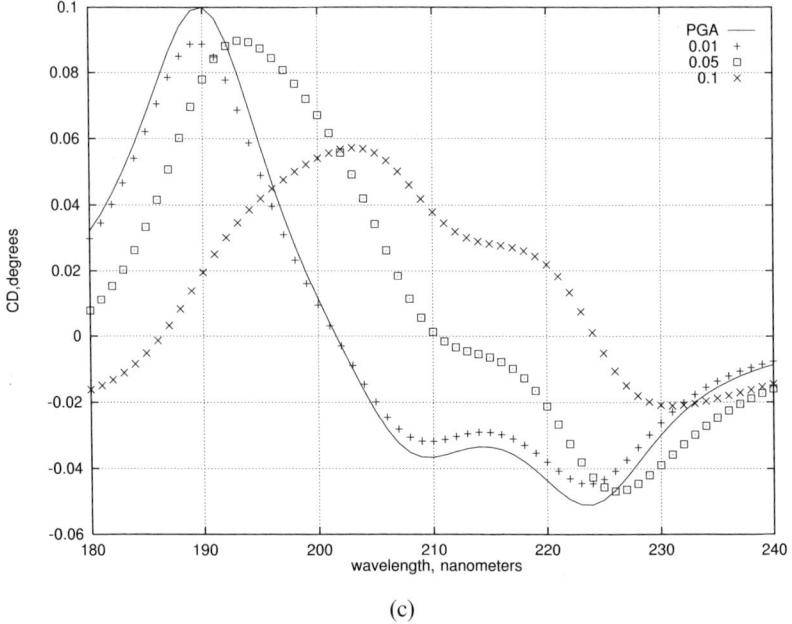

Figure 3.33 (*cont.*). (c) The same as in (b) but for the CD spectrum.

CD spectra of a particulate layer can look more like the ORD spectra of the solution. The physical description of such behaviour can be obtained in the framework of the anomalous diffraction approximation (Van de Hulst, 1981).

Note that, the function $\varphi(\lambda)$, $\psi(\lambda)$, and the elements of the scattering matrix can be used for monitoring the molecular structure of bio-particles during their lifetimes. This could be an important extension of general CD and ORD spectroscopy to the case of turbid media.

It should be pointed out that the elements of the extinction and phase matrices related to the optical activity phenomenon are very small. Thus, one can use the iteration method to solve the radiative transfer equation, where the first iteration is based on the ordinary radiative transfer equation in which the effects of chirality are neglected.

3.7.3 Rayleigh approximation

We present here simple approximations for the scattering and extinction matrices of Rayleigh particles. The elements of the amplitude matrix in the framework of the Rayleigh approximation are given by (Bohren and Huffman, 1983):

$$S_{11}(\theta) = 1.5 g_1 \gamma, \qquad S_{22}(\theta) = 1.5 g_1, \qquad S_{12}(\theta) = -S_{21}(\theta) = 1.5 c_1 (1 + \gamma),$$

where $\gamma = \cos\theta$ and θ is the scattering angle. Let us introduce the values $\xi = \frac{9}{4}|g_1|^2$, $\eta = \frac{9}{4}|c_1|^2$, and $q = \frac{9}{4}g_1 c_1^*$. In terms of these values: $i_1 = \xi\gamma^2$, $i_2 = \xi$, $i_3 = \eta(1+\gamma)^2$, $\mu = q\gamma(1+\gamma)$, $\nu = q(1+\gamma)$, and $\delta = \xi\gamma$. These equations can be used to find analytical expressions for the phase matrix elements in the framework of the Rayleigh approximation. Namely (the common multiplier w is omitted):

$$P_{11} = \frac{\xi}{2}(\gamma^2 + 1) + \eta(\gamma + 1)^2, \qquad P_{12} = P_{21} = \frac{\xi}{2}(\gamma^2 - 1),$$

$$P_{13} = -P_{31} = q'(\gamma^2 - 1), \qquad P_{14} = P_{41} = q''(\gamma + 1)^2,$$

$$P_{22} = \frac{\xi}{2}(\gamma^2 + 1) - \eta(\gamma + 1)^2, \qquad P_{23} = -P_{32} = q'(\gamma + 1)^2,$$

$$P_{24} = P_{42} = q''(\gamma^2 - 1), \qquad P_{33} = \xi\gamma - \eta(1+\gamma)^2,$$

$$P_{34} = -P_{43} = -\eta(1+\gamma)^2, \qquad P_{44} = \xi\gamma + \eta(1+\gamma)^2,$$

where $q' = \text{Re}(q)$, and $q'' = \text{Im}(q)$.

Let us consider these results in more detail. First of all it should be pointed out that $\eta \ll \xi$ in most cases. Thus it follows that, with a high accuracy for the diagonal elements,

$$P_{11} \approx P_{22} = \frac{\xi}{2}(1 + \cos^2\theta), \qquad P_{33} \approx P_{44} = \xi\cos\theta.$$

Elements P_{11} and P_{22} have a characteristic symmetric dependence on θ with the minimum at $\theta = \pi/2$. Elements P_{22} and P_{33} change from ξ to $-\xi$, while θ changes from 0 to π. The values of P_{33} and P_{44} almost vanish at $\theta = \pi/2$, where one can obtain:

$$P_{44}\left(\frac{\pi}{2}\right) = -P_{33}\left(\frac{\pi}{2}\right) = -P_{34}\left(\frac{\pi}{2}\right) = P_{43}\left(\frac{\pi}{2}\right) = \eta.$$

The dependence of values P_{12} on the scattering angle is similar to the case of isotropic spheres. However, the value P_{34} is not equal to zero for chiral Rayleigh spheres unlike the case of isotropic particles. The angular dependency of the elements P_{12}, P_{13}, and P_{24} are similar and differ by just a constant multiplier. The same conclusion is valid for the elements P_{34}, P_{14}, and P_{23}.

One can obtain for the elements of the normalized phase matrix $\hat{p}$:

$$b_1 = \frac{\gamma^2 - 1}{\gamma^2 + 1}, \qquad b_2 = -\frac{2\eta(1+\gamma)^2}{\xi(1+\gamma^2)},$$

$$b_3 = \frac{2q'(\gamma^2 - 1)}{\xi(\gamma^2 + 1)}, \qquad b_4 = \frac{2q'(1+\gamma)^2}{\xi(1+\gamma^2)},$$

$$b_5 = \frac{2q''(1+\gamma)^2}{\xi(1+\gamma^2)}, \qquad b_6 = \frac{2q''(\gamma^2 - 1)}{\xi(1+\gamma^2)},$$

$$a_2 = 1, \qquad a_3 = a_4 = \frac{2\gamma}{1+\gamma^2},$$

where we have neglected small terms in the diagonal elements. One can see that the angular dependence of elements b_1, b_3, and b_6 is similar. The same is true for elements b_2, b_4, and b_5. They do not depend on the size of the particles. The correspondence between this approximation and exact results is shown in Figure 3.31.

One can obtain for values of the CD and ORD:

$$\varphi = \frac{q''\gamma(1+\gamma)}{\xi\gamma^2 + \eta(1+\gamma)^2}, \qquad \psi = \frac{q'\gamma(1+\gamma)}{\xi\gamma^2 - \eta(1+\gamma)^2}.$$

Note that the coefficients g_1 and c_1 in the framework of the Rayleigh approximation are:

$$g_1 = -\frac{2i}{3} f_1(m) x^3,$$

$$c_1 = \frac{(m_L - m_R)}{f_2(m)} x^3,$$

where

$$f_1(m) = \frac{(3m^2 - 1)\bar{m} - 2m}{(3m^2 + 2)\bar{m} + 4m},$$

$$f_2(m) = 3m\bar{m} + 2\frac{\bar{m}}{m} + 4.$$

Values of m_L and m_R are close to each other in most cases and $m \approx \bar{m}$. It follows that in this approximation:

$$g_1 = -\frac{2i}{3} \frac{m^2 - 1}{m^2 + 2} x^3,$$

$$c_1 = \frac{m_L - m_R}{3(m^2 + 2)} x^3.$$

3.8 FURTHER READING

Aden, A. L. and Kerker, M. (1951) Scattering of electromagnetic waves from two concentric spheres. *J. Appl. Phys.*, **22**, 1242–1246.

Arnott, W. P. and Marston, P. L. (1988) Optical glory of small freely rising gas bubbles in water: observed and computed cross-polarized backscattering patterns. *J. Opt. Soc. America*, A **5**, 496–506.

Asano, S. and Yamamoto, G. (1975) Light scattering by a spheroidal particle. *Appl. Opt.*, **14**, 29–49.

Babenko, V. A., Astafyeva, L. G., and Kuzmin, V. N. (2003) *Electromagnetic Scattering in Disperse Inhomogeneous and Anisotropic Particles*. Springer–Praxis, Chichester, UK (in press).

Baran, A. (2002) A scattering phase function for ice cloud: Tests of applicability using aircraft and satellite multi-angle multi-wavelength radiance measurements of cirrus. *Quart. J. Royal Meteorol. Soc.*, **127**, 2395–2416.

Barber, P. W. and Yeh, C. (1975) Scattering of electromagnetic waves by arbitrary shaped dielectric bodies. *Appl. Opt.*, **14**, 2864–2872.

Barber, P. W. and Hill, S. C. (1990) *Light Scattering by Particles: Computational Methods.* World Scientific, Singapore.

Barkey, B. and Liou, K. N. (2001) Polar nephelometer for light-scattering measurements of ice crystals. *Optics Letters*, **26**, 232–234.

Beardsley, G. F. Jr. (1968) Mueller scattering matrix of sea water. *J. Opt. Soc. America*, **58**, 52–57.

Bohren, C. F. (1974) Light scattering by an optically active sphere. *Chem. Phys. Lett.*, **29**, 458–462.

Bohren, C. F. (1975a) *Light Scattering by Optically Active Particles.* PhD thesis, University of Arizona, Tuscon, AZ.

Bohren, C. F. (1975b) Scattering of electromagnetic waves by an optically active spherical shell. *J. Chem. Phys.*, **62**, 1566–1571.

Bohren, C. F. (1978) Scattering of electromagnetic waves by an optically active cylinder. *J. Colloid Interface Sci.*, **66**, 105–109.

Bohren, C. F. and Huffman, D. R. (1983) *Absorption and Scattering of Light by Small Particles.* Wiley, New York.

Born, M. and Wolf, E. (1999) *Principles of Optics.* Cambridge University Press, Cambridge, UK.

Breon, F.-M. and Goloub, P. (1998) Cloud droplet effective radius from space-borne polarization measurements. *Geophys. Res. Let.*, **25**, 1879–1883.

Brosseau, C. and Bicout, D. (1994) Entropy production in multiple scattering of light in a spatially random medium. *Physical Review E*, **50**, 4997–5005.

Card, J. B. A. and Jones, A. R. (1999) An investigation of the potential of light scattering for the characterization of irregular particles. *J. Phys., D, Appl. Phys.*, **32**, 2466–2474.

Chylek, P., Grams, G. W., and Pinnick, R. G. (1976) Light scattering by irregular randomly oriented particles. *Science*, **193**, 480–482.

Cooray, M. R. F. and Ciric, I. R. (1993) Wave scattering by a chiral spheroid. *J. Opt. Soc. America*, **A10**, 1197–1203.

Dau-Sing, W. and Barber, P. W. (1979) Scattering by inhomogeneous nonspherical objects. *Appl. Opt.*, **18**, 1190–1197.

Davis, G. E. (1952) Scattering of light by an air bubble in water. *J. Opt. Soc. America*, **45**, 572–581.

Deirmendjian, D. (1969) *Light Scattering by Spherical Polydispersions*, Elsevier, New York.

Draine, B. T. and Flatau, P. J. (1994) Discrete-dipole approximation for scattering calculations. *J. Opt. Soc. Am.*, **A11**, 1491–1499.

Drossart, P. (1990) A statistical model for the scattering by irregular particles. *Astrophys. J.*, **361**, L39–L42.

Dugin, V. P. and Mirumyants, S. O. (1976) The light scattering matrices of artificial crystalline clouds. *Izv. Acad. Sci. USSR, Atmos. Oceanic Phys.*, **12**, 988–991.

Farafonov, V. G. (2000) Light scattering by multilayered ellipsoids in the Rayleigh approximation. *Opt. and Spectr.*, **88**, 441–443.

Fedorov, F. I. (1976) *Theory of Gyrotropy.* Nauka and Tekhnika, Minsk.

Garett, T. J., Hobbs, P. V. and Gerber, H. (2001) Shortwave, single-scattering properties of arctic ice clouds. *J. Geophys. Res.*, **106**(14), 15155–15172.

Gorchakov, G. I. (1966) Light scattering matrices of atmospheric air in a boundary layer. *Izvestiya, Atmospheric and Oceanic Physics*, **2**, 595–605.

Gorchakov, G. I. (1971) On the degree of the polarization coherence of light scattered by atmospheric air. *Izvestiya, Atmospheric and Oceanic Physics*, **7**, 224–229.

Grandy, W. T. (2000) *Scattering of Waves from Large Spheres*. Cambridge University Press, Cambridge, UK.

Hanel, G. (1972) Computation of the extinction of visible radiation by atmospheric aerosol particles as a function of the relative humidity, based upon measured properties. *J. Aerosol Sci.*, **3**, 337–386.

Hansen, J. E. and Travis, L. D. (1974) Light scattering in planetary atmospheres. *Space Sci. Rev.*, **16**, 527–610.

Hovenier, J. W., Van de Hulst, H. C., and van der Mee, C. V. M. (1986) Conditions for the elements of the scattering matrix. *Astron. and Astrophys.*, **157**, 301–310.

Ishimaru, A. and Yeh, C. W. (1984) Matrix representations of the vector radiative transfer theory for randomly distributed nonspherical particles. *J. Opt. Soc. America*, **A1**, 359–364.

Jourdan, O., Oshchepkov, S., Gayet, J.-F., Shcherbakov, V. and Isaka, H. (2003). Statistical analysis of cloud light scattering and microphysical properties obtained from airborne measurements. *J. Geophys. Res. D*, **108**, 10.1029/2002JD002723.

Kahnert, F. M., Stamnes, J. J. and Stamnes, K. (2002) Using simple particle shapes to model the Stokes scattering matrix of ensembles of wavelength-sized particles with complex shape: Possibilities and limitations. *J. Quant. Spectr. and Rad. Transfer*, **74**, 167–182.

Kerker, M. (1969) *The Scattering of Light and Other Electromagnetic Radiation*. Academic Press, New York.

Kerker, M. (ed) (1988) *Selected Papers on Light Scattering*, SPIE Proceedings, v. 951.

Khlebtsov, N. G., Melnikov, A. G. and Bogatyrev, V. A. (1991) The linear dichroism and birefringence of colloidal dispersions: Approximate and exact approaches. *J. Colloid Interface Sci.*, **146**, 463–478.

Khlebtsov, N. G., Melnikov, A. G., Bogatyrev, V. A. and Sirota, A. I. (1999) Electrooptic effects in dilute suspensions of bacterial cells and fractal aggregates. *J. Quant. Spectr. Radiat. Transfer*, **63**, 469–478.

Khlebtsov, N. G. and Bogatyrev, V. A. (2001) Structural anisotropy of fractal clusters and its manifestation in electrooptical effects. *Colloid. J.*, **63**, 528–537.

Khlebtsov, N. G., Maksimova, I. L., Tuchin, V. V. and Wang, L. (2002) Introduction to light scattering by biological objects. *Handbook of Optical Biomedical Diagnostics* (edited by V. V. Tuchun, Ch. 1, pp. 31–167). Bellingham, Washington.

Kokhanovsky, A. A. (1999) Radiative transfer in chiral random media. *Phys. Rev. E*, **60**, 4899–4905.

Kokhanovsky, A. A. (2001) *Light Scattering Media Optics: Problems and Solutions*. Springer–Praxis, Chichester, UK.

Kuik, F. (1992) *Single Scattering of Light by Ensembles of Particles with various shapes*. Amsterdam Free University, Enschede, The Netherlands.

Kuik, F., de Haan, J. F. and Hovenier, J. W. (1992) Benchmark results for single scattering by spheroids. *J. Quant. Spectr. Radiative Transfer*, **47**, 477–489.

Lacoste, D., van Tiggelen, B. A., Rikken, G. L. J. A. and Sparenberg, A. (1998) Optics of a Faraday-active sphere Mie sphere. *J. Opt. Soc. America*, A15, 1636–1642.

Lacoste, D. (1999) *Diffusion de la lumiere dans les milieux magneto-optiques et chiraux*. PhD thesis, Université Joseph Fourier-Grenoble I, Grenoble, France.

Liou, K.-N., Cai, Q., Barber, P. W. and Hill, S. C. (1983) Scattering phase matrix comparison for randomly oriented hexagonal cylinders and spheroids. *Appl. Opt.*, **22**, 1684–1687.

Liou, K. N. (2002) *An Introduction to Atmospheric Radiation*. Academic Press, New York.

Macke, A. (1993) Scattering of light by polyhedral ice crystals. *Appl. Opt.*, **32**, 2780–2788.
Macke, A. (1994) Modellierung der Optischen Eigenschaften von Cirruswolken. PhD thesis, University of Hamburg.
Macke, A., Mishchenko, M. I., Muinonen, K. and Carlson, B. E. (1995) Scattering of light by large nonspherical particles: Ray-tracing approximation versus T-matrix method. *Opt. Lett.*, **20**, 1934–1936.
Macke, A. and Mischenko, M. (1996) Applicability of regular particle shapes in light scattering calculations for atmospheric ice particles. *Appl. Optics*, **35**, 4291–4296.
Macke, A., Mueller, J. and Raschke, E. (1996) Scattering properties of atmospheric ice crystals. *J. Atmos. Sci.*, **53**, 281–297.
Mandelbrot, B. B. (1977) *The Fractal Geometry of Nature*. W. H. Freeman and Company, New York.
Mie, G. (1908) Beiträge zur Optik trüber Medien speziell kolloidaler Metallösungen. *Ann. Phys.*, **25**, 377–445.
Mishchenko, M. I., Travis, L. D. and Lacis, A. A. (2002) *Scattering, Absorption, and Emission of Light by Small Particles*. Cambridge University Press, Cambridge, UK.
Muinonen, K., Nousiainen, T., Fast, P., Lumme, K. and Peltoniemi, J. I. (1996) Light scattering by Gaussian random particles: Ray optics approximation. *J. Quant. Spectr. and Rad. Transfer*, **55**, 577–601.
Nevitt, T. J. and Bohren, C. F. (1984) Infrared backscattering by irregularly shaped particles: A statistical approach. *J. Climate App. Meteorol.*, **23**, 1342–1349.
Newton, R. (1982) *Scattering Theory of Waves and Particles*. Springer-Verlag, New York.
Nousiainen, T. and Muinonen, K. (1999) Light scattering by Gaussian, randomly oscillating raindrops. *J. Quant. Spectr. Rad. Transfer*, **63**, 643–666.
Nousiainen, T. (2000) Scattering of light by raindrops with single-mode oscillations. *J. Atmos. Sci.*, **57**, 789–802.
Osborn, J. A. (1945) Demagnetizing factors for the general ellipsoid. *Phys. Rev.*, **67**, 351–357.
Perrin, F. (1942) Polarization of light scattered by isotropic opalescent media. *J. Chem. Phys.*, **10**, 415–427.
Prishivalko, A. P. and Astafieva, L. G. (1975) *Absorption, Scattering, and Extinction of Light by Water-Covered Atmospheric Particles*. Institute of Physics, Minsk.
Prishivalko, A. P. (1976) Uptake of water by particles and matrices of light scattering by atmospheric aerosol. Institute of Physics, Minsk.
Prishivalko, A. P., Babenko, V. A. and Kuzmin, V. N. (1984) *Scattering and Absorption of Light by Inhomogeneous and Anisotropic Spherical Particles*. Nauka and Tekhnika, Minsk.
Pulfrich, C. (1888) Uber eine dem Regenbogen verwandte Erscheinung der Toatalreflexion. *Ann. Phys. Chem.*, **33**, 209–212.
Shifrin, K. S. (1951) *Scattering of Light in a Turbid Media*. Gostekhteorizdat, Moscow. (English translation: NASA Tech. Trans. TT F-447, 1968, NASA, Washington, DC).
Shifrin, K. S. (1952) Light scattering on two-layered particles. *Izvestiya, Geophysics*, **2**, 15–21.
Shifrin, K. S. and Mikulinskii, I. A, (1982) Light scattering by a system of particles in Rayleigh–Gans approximation. *Opt. Spectr.*, **52**, 359–366.
Shifrin, K. S., Shifrin, Ya, S. and Mikulinski, I. A. (1984) Light scattering by an ensemble of large particles of arbitrary shape. *Doklady Akad. Nauk SSSR*, **277**, 582–585.
Shifrin, K. S. (1988) *Introduction to Ocean Optics*. Gidrometeoizdat, Leningrad.
Shuerman, D. (ed) (1983) *Light Scattering by Irregularly Shaped Particles*. Plenum Press, New York.

Sid'ko, F. Ya., Lopatin, V. N. and Paramonov, L. E. (1990) *Polarization Characteristics of Biological Suspensions*. Nauka, Novosibirsk.

Silva, A. M., Bugalho, M. L., Costa, M. J., von Hoyningen-Huene, W., Schmidt, T. and Heintzenberg, J. (2002) Aerosol optical properties from columnar data during the second aerosol characterization experiment on the south coast of Portugal. *J. Geophys. Res. D*, **107**, 10.1029/2002JD002196.

Stammes, P. (1989) *Light Scattering Properties of Aerosols and the Radiation inside a Planetary Atmosphere*. Amsterdam Free University, Enschede, The Netherlands.

Takano, Y. and Liou, K.-N. (1989) Solar radiative transfer in cirrus clouds. 1: Single scattering and optical properties of hexagonal ice crystals. *J. Atmos. Sci.*, **46**, 3–19.

Takano, Y. and Liou, K. N. (1995) Radiative transfer in Cirrus clouds. Part III: Light scattering by irregular ice crystals. *J. Atmos. Sci.*, **52**, 818–837.

Tricker, R. A. R. (1970) *Introduction to Meteorological Optics*. Elsevier, New York.

Tyler, J. E. (1961) Measurements of scattering properties of hydrosols. *J. Opt. Soc. America*, **51**, 1289–1293.

Van de Hulst, H. C. (1981) *Light Scattering by Small Particles*. Dover, New York.

Volten, H. (2001) *Light Scattering by Small Planetary Particles*. PhD thesis, Amsterdam Free University, Enschede, The Netherlands.

Volten, H., Muñoz, O., Rol, E., de Haan, J. F., Vassen, W., Hovenier, J. W. et al. (2001) Scattering matrices of mineral particles at 441.6 and 632.2 nm. *J. Geophys. Res.*, **D106**, 17375–17401.

Von Hoyningen-Huene, W. and Posse, P. (1997) Nonsphericity of aerosol particles and their contribution to radiative forcing. *J. Quant. Spectr. Rad. Transfer*, **57**, 651–668.

Voss, K. J. and Fry, E. S. (1984) Measurement of the Mueller matrix for ocean water. *Appl. Opt.*, **23**, 4427–4439.

Wiscombe, W. J. and Mugnai, A. (1986) *Single Scattering from Nonspherical Chebeshev Particles: A Compendium of Calculations* (NASA Ref. Pub. 1157). NASA Goddard Space Flight Center, H. C. Greenbelt, MD.

Wriedt, T. (2002) Using the T-matrix light scattering computations by non-axisymmetric particles: Superellipsoids and realistic shapes. *Part. Part. Charact.*, **19**, 256–268.

Yang, P. and Liou, K. N. (1995) Light scattering by hexagonal ice crystals: Comparison of finite-difference time domain and geometric optics models. *J. Opt. Soc. Am.*, **12**, 162–176.

Yang, P. and Liou, K. N. (1996a) Finite-difference time domain method for light scattering by small ice crystals in three-dimensional space. *J. Opt. Soc. Am.*, **A13**, 2072–2085.

Yang, P. and Liou, K. N. (1996b) Geometric-optics-integral-equation method for light scattering by nonspherical ice crystalls. *Appl. Opt.*, **35**, 6568–6584.

Zaneveld, J. R. and Pack, H. (1973) Method for the determination of the index of refraction of particles in the ocean. *J. Opt. Soc. America*, **63**, 321–324.

Zhang, J. and Xu, L. (1995) Light scattering by absorbing hexagonal ice crystals in cirrus clouds. *Appl. Opt.*, **34**, 5867–5874.

4

Environmental polarimetry

4.1 CLOUDS

4.1.1 Water clouds

Let us now consider the polarization characteristics of light in natural media, starting with the case of clouds. The intensity and polarization characteristics of singly scattered light for spherical polydispersions of water droplets were considered in Chapter 3. In particular, it was shown that the degree of polarization of singly scattered light could reach 80% in the vicinity of the rainbow angle (see Figure 3.1(b)). Such strong polarization, however, is not observed for terrestrial clouds due to the influence of the effects of multiple light scattering. The contribution of singly scattered photons to the reflected light field is small in comparison with that of multiply scattered photons. Multiple light scattering does not produce additional polarization of singly scattered beams, but only diminishes the polarization which arises from single light scattering. Therefore, the degree of polarization, obtained in the single scattering gives the upper border for a given type of scatterer. For realistic clouds, the polarization of both reflected and transmitted light is much lower. We present results of calculations of the degree of polarization p_e for weakly absorbing cloudy media, having different optical thickness τ, in Figure 4.1. They were obtained by solution of the vector radiative transfer equation (Hansen, 1971). The angular distribution of reflection function R (intensity in Figure 4.1) is also given.

We see the positive peak in polarization for the rainbow angle. The degree of polarization in the peak is in the range 5–17% for cloud optical thickness in the range 4–30 and effective size of droplets 6 μm (compare with Figure 3.1(b)). The degree of polarization is somewhat lower outside the rainbow peak. It vanishes in the exact forward/backward directions and around the observation angles 25 and 80°. The scattering angles at these observation angles are 155 and 100° respectively. These

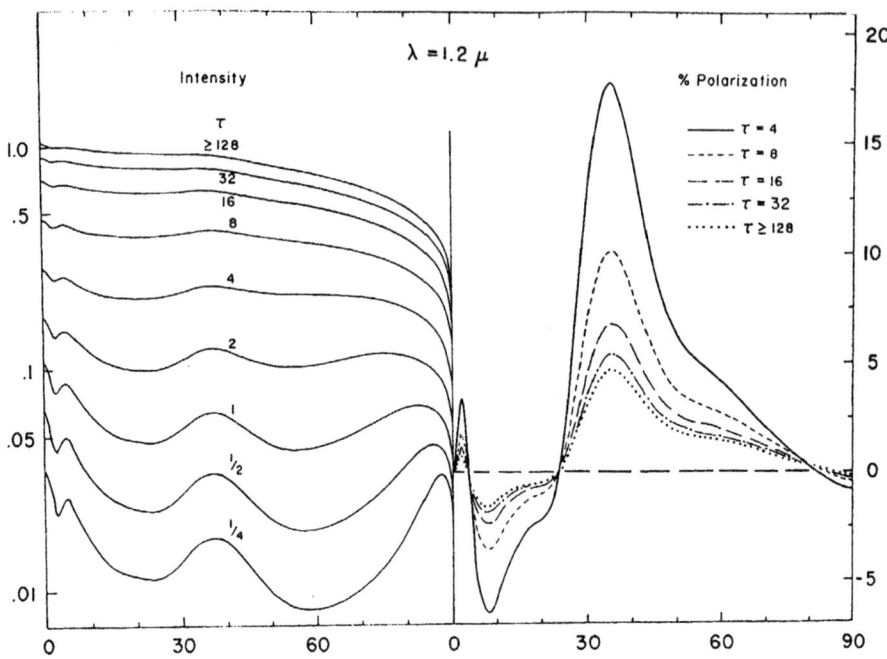

Figure 4.1. The dependence of the reflection function and the degree of polarization (%) of reflected light on the zenith observation angle with the sun overhead for selected values of τ. It is assumed that water droplets in a cloud are characterized by the particle size distribution (PSD) (1.5) with $a_0 = 4\,\mu\text{m}$ and $\mu = 6$. The wavelength is equal to $1.2\,\mu\text{m}$ (Hansen, 1971).

are the angles, where the polarization of singly scattered light vanishes (see Figure 3.1(b)). It is interesting to see these manifestations of light interaction with a single droplet in a multiple light scattering case.

The polarization near the rainbow angle oscillates from a negative to a positive value, which is also due to special features of the degree of polarization for singly scattered light (compare with Figure 3.1(b)). The case $\tau = 128$ roughly corresponds to a semi-infinite cloud. Note that the angular dependencies of the degree of polarization for all optical thicknesses are similar. They can be obtained from the dependence $p_l(\theta)$ at $\tau = 128$ by use of a multiplication factor, introduced in Chapter 2 (see Equation 2.89). It approximately equals the inverse spherical albedo of a cloud at $\omega_0 = 1$. Functions $R(\vartheta)$ and $p_l(\vartheta)$ at $\tau = 32$ and various sizes of particles are given in Figure 4.2. We see that the peak of polarization increases and moves to smaller scattering angles $\theta = \pi - \vartheta$ with the size of droplets. Therefore, it is possible to obtain the size of droplets from the degree of polarization measurements (Breon and Goloub, 1998; Kokhanovsky and Weichert, 2002). We emphasize that the role of multiple light scattering is to diminish polarization, preserving main angular features, which are seen in the singly scattered light (see Figure 3.1(b)).

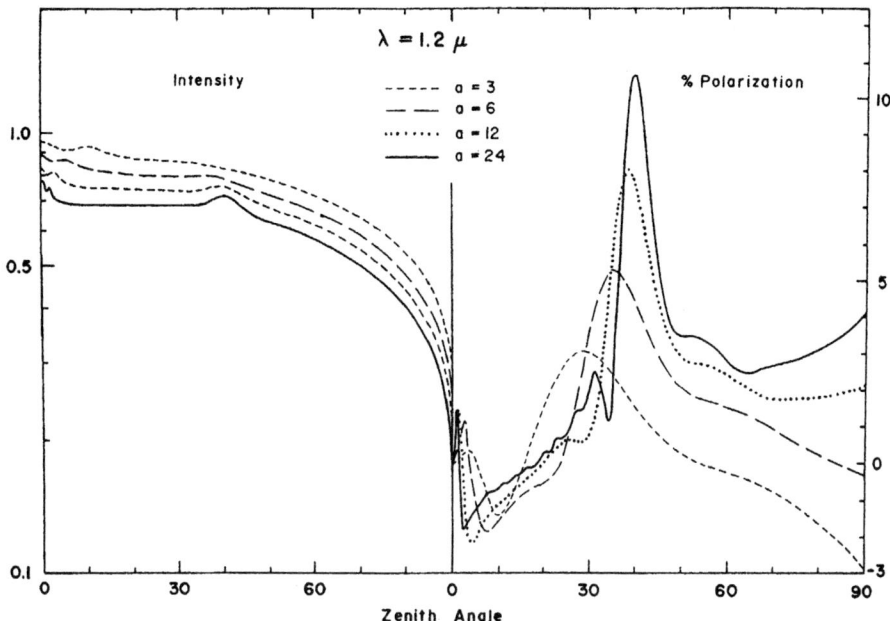

Figure 4.2. The same as in Figure 4.1 but for the fixed cloud optical thickness equal to 32 and different effective radii equal to 3, 6, 12, and 24 μm (Hansen, 1971).

The functions $R(\vartheta)$ and $p_l(\vartheta)$ in the visible spectrum are similar to those considered above. They differ considerably, however, for larger wavelengths. This is due to high absorption of electromagnetic waves by water there. Characteristic examples are given in Figure 4.3 ($\lambda = 2.25\,\mu m$) and Figure 4.4 ($\lambda = 3.4\,\mu m$). We see that absorption leads to a decrease in reflected light intensity. The case of $\tau = 4$ can be considered as a semi-infinite layer at $\lambda = 3.4\,\mu m$, where absorption of electromagnetic waves by water is high.

Summarizing, we underline that polarization characteristics of clouds are highly dependent on the wavelength. In the visible, where absorption is negligible, multiple light scattering greatly reduces the degree of polarization of singly scattered light, preserving however, the angular variation pattern. This pattern is dependent on the refractive index of the particles. Therefore, the angular variation of $p_l(\theta)$ (e.g., the position of rainbow and zero points) can be used to identify the chemical composition of clouds (if they differ from water, e.g., on other planets). The degree of polarization is usually below 20% in the visible range of the electromagnetic spectrum when the sun is overhead. In the infrared range, where absorption of light by droplets increases, we have smaller and smaller fractions of multiply scattered light. In the limit of infinite absorption ($\chi a_{ef} \to \infty$), the degree of polarization of reflected light is completely determined by the external reflection of light from droplets. This external reflection does not depend on the size and shape for randomly oriented particles, and can be used to retrieve the complex refractive index of cloud particles (not necessarily spheres).

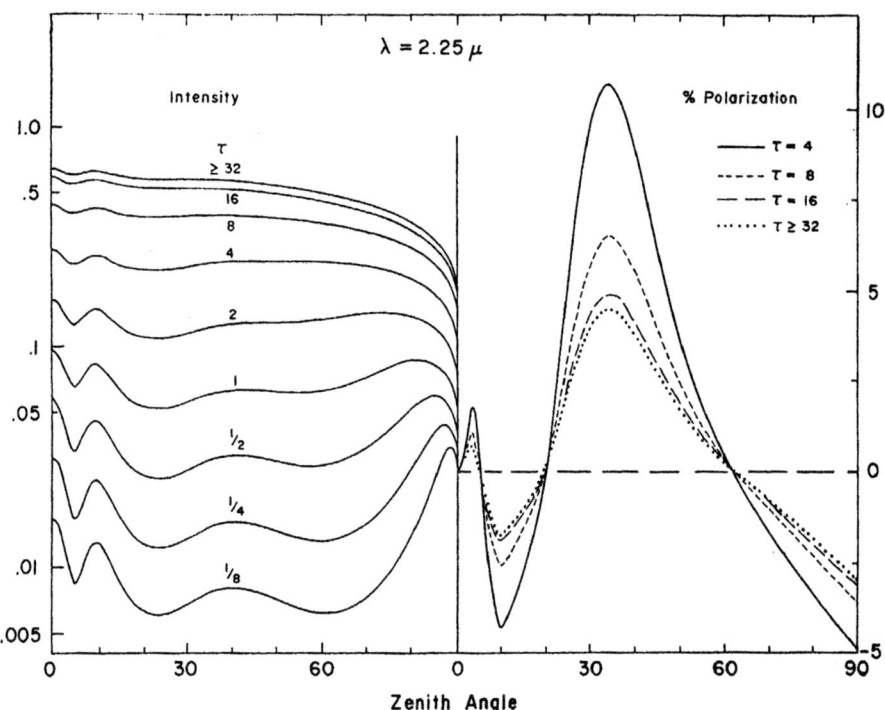

Figure 4.3. The same as Figure 4.1 except $\lambda = 2.25\,\mu m$ (Hansen, 1971).

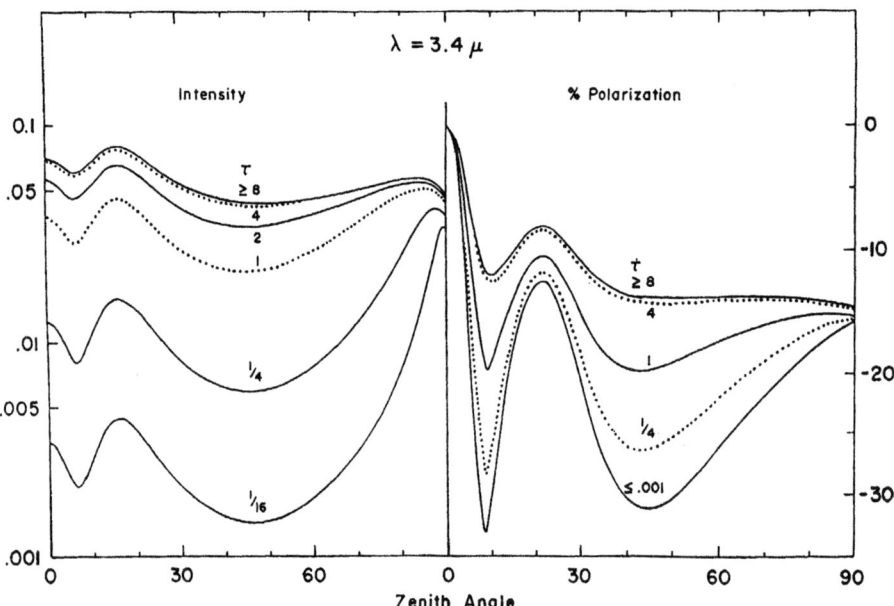

Figure 4.4. The same as in Figure 4.1 except $\lambda = 3.4\,\mu m$ (Hansen, 1971).

4.1.2 Ice clouds

The degree of polarization of light reflected from ice clouds is much lower than that of water clouds. This is due to the absence of rainbow and glory scattering for crystalline clouds. Also the shape of the particles is very irregular and chaotic in ice clouds. Therefore, they are not able to produce high polarization of incident unpolarized solar light (see Figure 3.24(a)). Generally, multiple light scattering reduces the degree of polarization of cloudy media. This is the reason for almost zero polarization of light transmitted by ice and water clouds. The degree of polarization in transmitted light can increase due to reflection of partially polarized light beams, coming from the ground surface and aerosols underneath clouds.

The value of p_l is already quite low for single scattering (see Figure 4.5) by crystals and it is further reduced due to multiple scattering effects. The dependence of the degree of polarization on scattering angle for optically thin cirrus clouds (with $\tau \approx 3$) is given in Figure 4.6 for various wavelengths. We see that p_l is extremely low in the visible range of the electromagnetic spectrum. It generally increases with the wavelength.

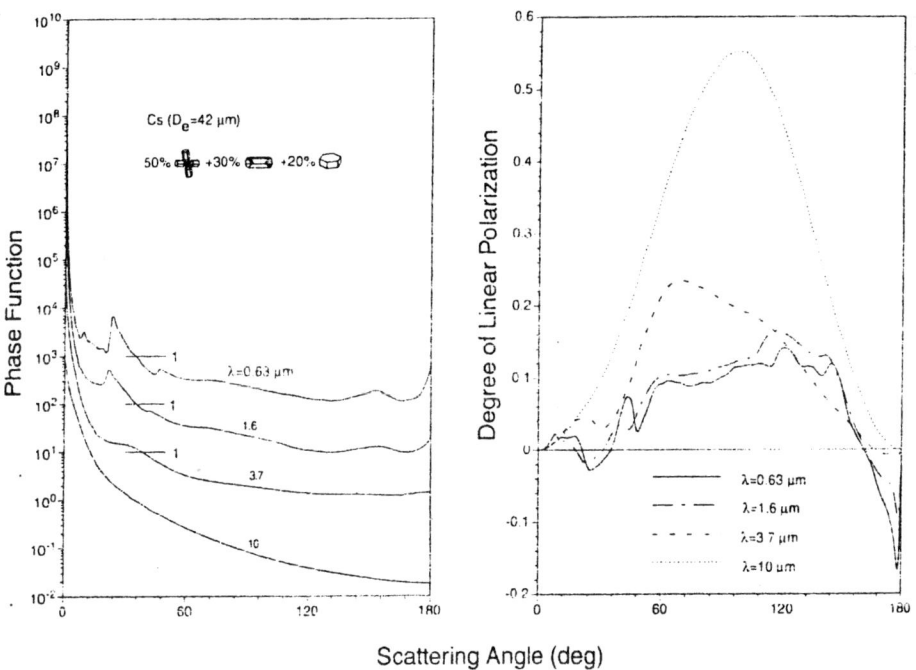

Figure 4.5. Phase function and degree of linear polarization patterns for a typical cirrostratus with a mean effective size of about 42 μm composed of 50% bullet rosettes/aggregates, 30% hollow columns, and 20% plates – a shape model based on replicator and optical probe measurement. Four remote sensing wavelengths are displayed. In the phase function, the vertical scale applies to the lowest curve; the upper curves are displayed upward by a factor of 10 (Liou et al., 2002).

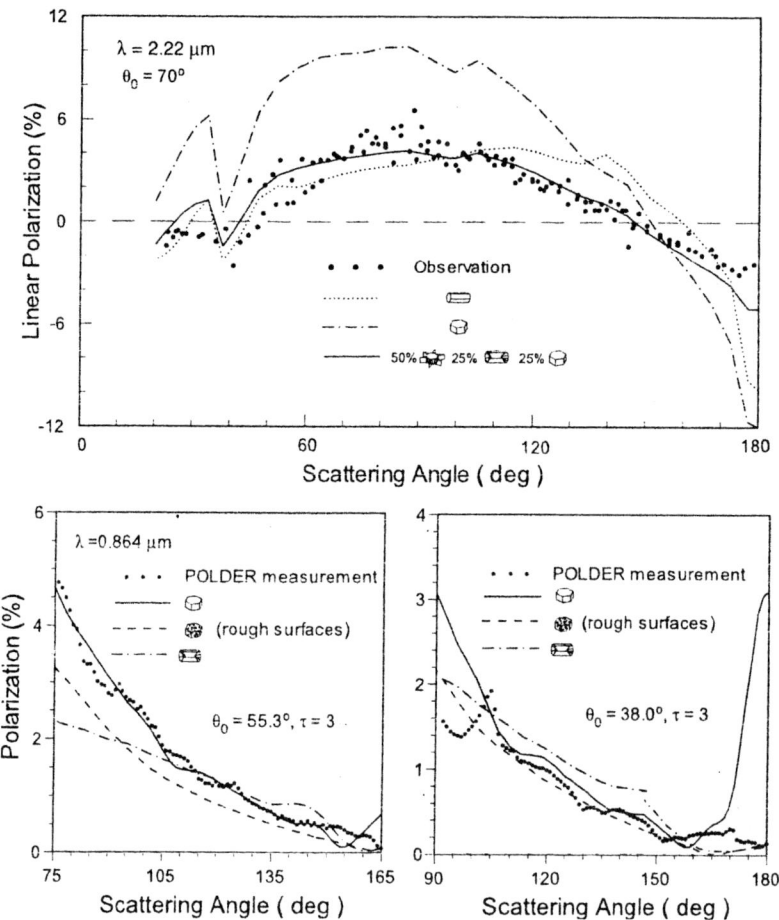

Figure 4.6. The upper panel shows linear polarization of sunlight reflected from cirrus clouds measured at 2.22 μm (Coffeen, 1979). The lower panels display full polarization observed from the polarimeter POLDER at 0.864 μm (Chepfer et al., 1998). The theoretical results in the lower panel are computed for hollow columns, plates, and ice particles with rough surfaces using the best-fit optical depth 3 (Liou et al., 2002). Theoretical results in the upper panel are obtained using single scattering calculations for hexagonal plates and columns and a combination of regular and irregular ice crystals. θ_0 is the solar zenith angle.

The angular dependence of the degree of polarization of crystalline clouds differs from the case of liquid clouds. This can be used for the discrimination of ice/water clouds in satellite imagery.

The measured degree of polarization at $\lambda = 0.86$ μm and $\lambda = 2.2$ μm (see also Figure 4.3) is smaller than 6% for all scattering angles and can be well represented by a mixture of 50% plates with attachments, and an equal proportion of hexagonal plates and hollow columns (see upper plate in Figure 4.6). Again this does not necessarily mean that ice clouds are predominantly composed of these particles. It follows from measurements (see the lower plate in Figure 4.6) that the degree of polarization typically decreases with angle in the backward hemisphere.

Another example of ice clouds is given by noctilucent or polar mesospheric clouds, which exist at a very high altitude (approximately 80 km as compared to just 0.5–10 km for tropospheric clouds). It was found that ice crystals in these clouds are very small in comparison to the wavelength of the visible light. Therefore, their scattering diagram and polarization characteristics are close to the Rayleigh case (but with depolarization effects due to the non-sphericity of particles as described in the previous chapter).

4.2 AEROSOLS

The average optical thickness of aerosols in the visible range of the electromagnetic spectrum is close to 0.2 with small values around 0.05 for remote areas (e.g., over some parts of the ocean). This means that their polarization characteristics are largely (but not exclusively) determined by that of singly scattered light. There are some exceptions, of course, like dust outbreaks from African and Asian deserts, where the optical thickness can reach 3–5.

The World Meteorological Organization classifies aerosols in four major classes. They are oceanic aerosols, dust aerosols, soot, and water-soluble aerosols. There are also other types of aerosols (e.g., organic aerosols, which originate, e.g., from the gas phase over forests). However, we concentrate on the polarization characteristics of these four types of aerosols. Their microphysical characteristics are summarized in Table 4.1.

Phase matrix elements for all four aerosols types are given in Figures 4.7 and 4.8. Aerosols over different areas can be considered as mixtures of these four basic types.

Table 4.1. Parameters of particle size distributions of selected aerosol types (WCP-112, 1986).

Aerosol type	PSD	a_{ef} (µm)	Δ
Water-soluble (w)	Lognormal $\sigma = 1.09527$ $a_m = 0.05$ µm	0.1	1.52
Dust (d)	Lognormal $\sigma = 1.09527$ $a_m = 0.5$ µm	10.0	1.52
Soot (s)	Lognormal $\sigma = 0.69317$ $a_m = 0.0118$ µm	0.04	0.79
Oceanic (o)	Lognormal $\sigma = 0.92028$ $a_m = 0.3$ µm	2.5	1.15

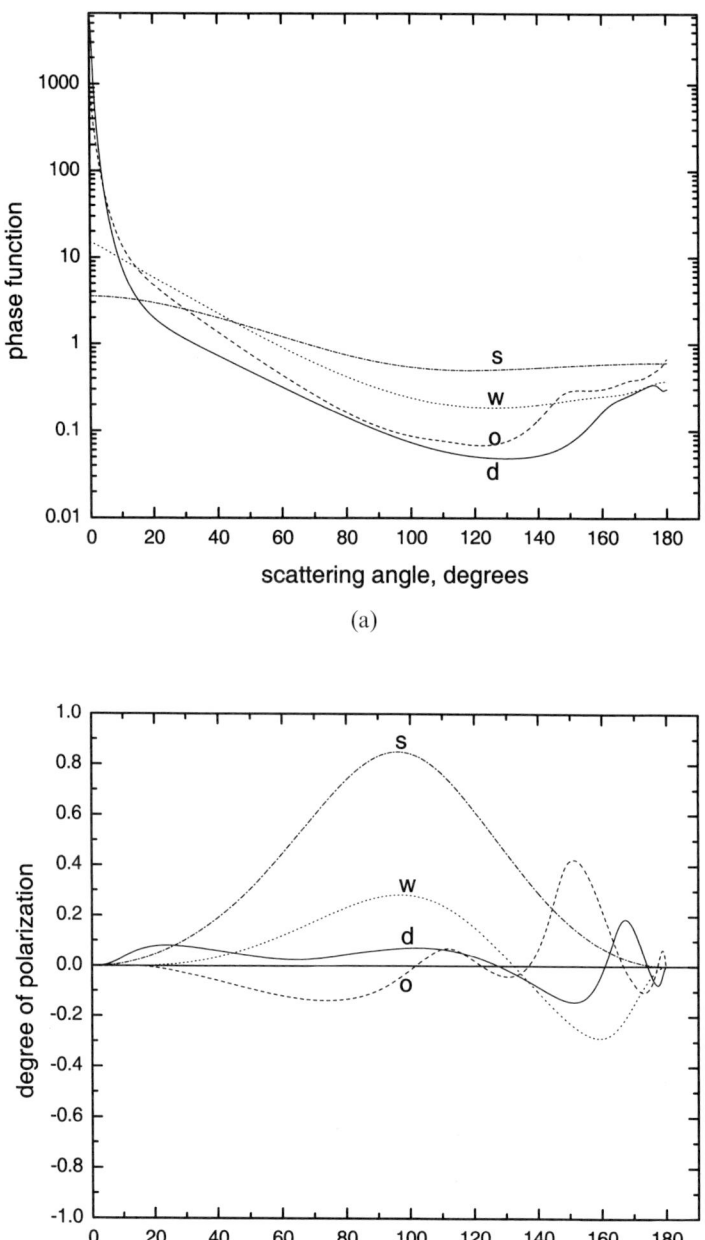

Figure 4.7. (a) The phase function of the water-soluble (w), soot (s), oceanic (o), and dust (d) aerosols at the wavelength 0.55 μm; (b) the same as in (a) except for the degree of polarization p_l.

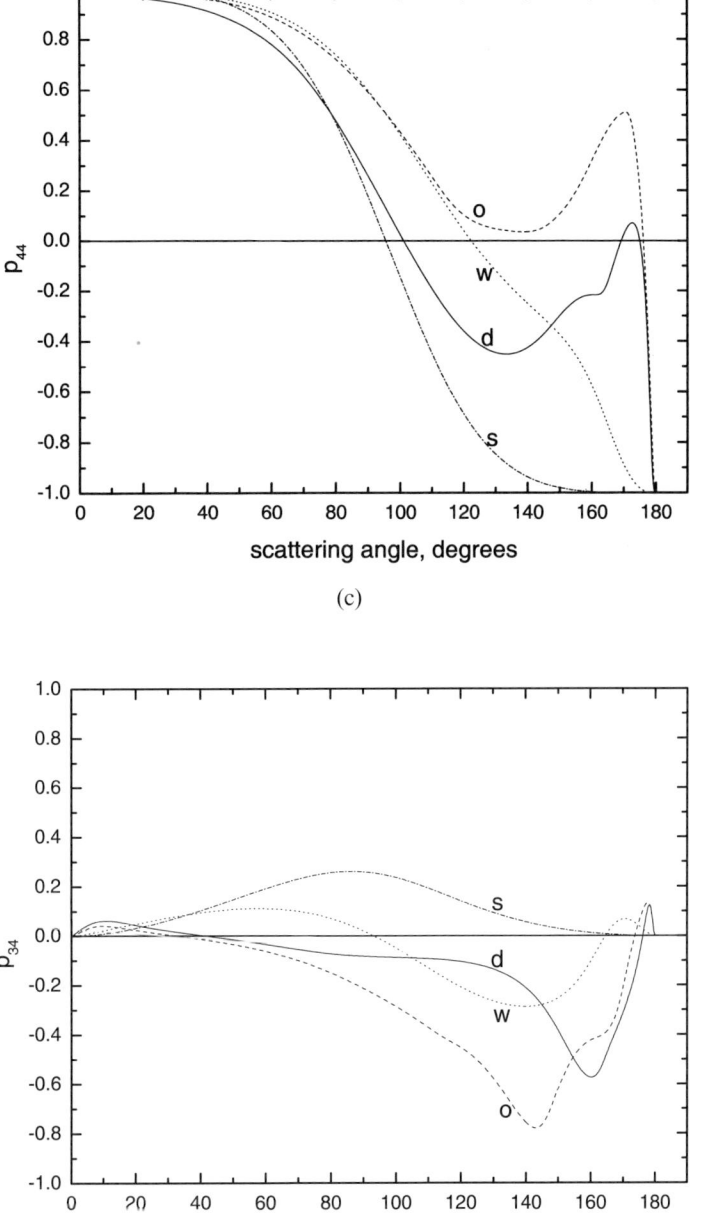

Figure 4.7 (*cont.*). (c) The same as in (a) except for p_{44}; (d) the same as in (a) except for p_{34}.

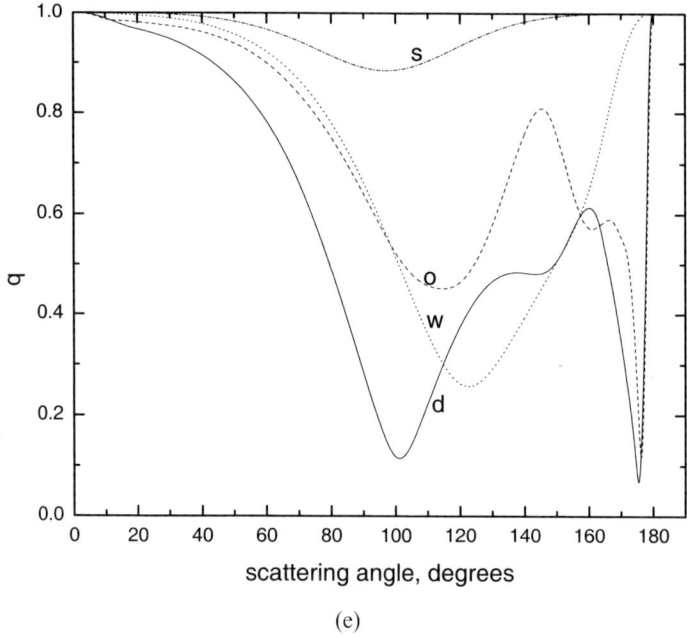

(e)

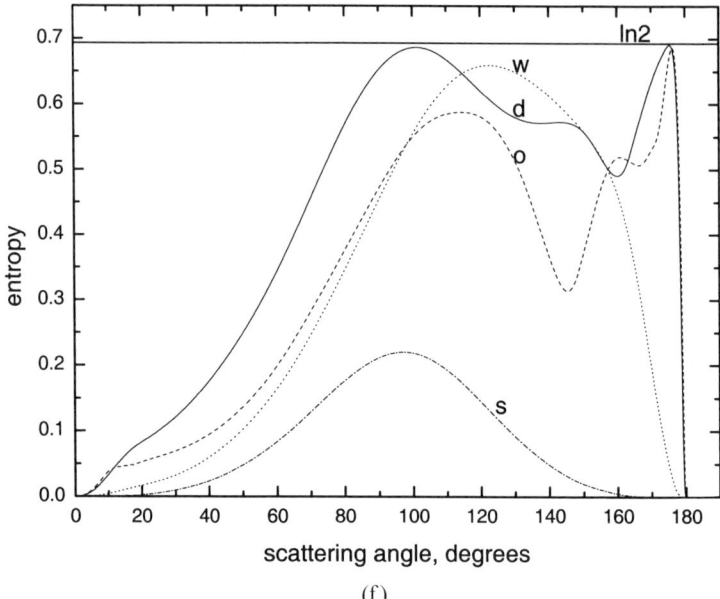

(f)

Figure 4.7 (*cont.*). (e) The same as in (a) except for q; (f) the same as in (a) except for the entropy s.

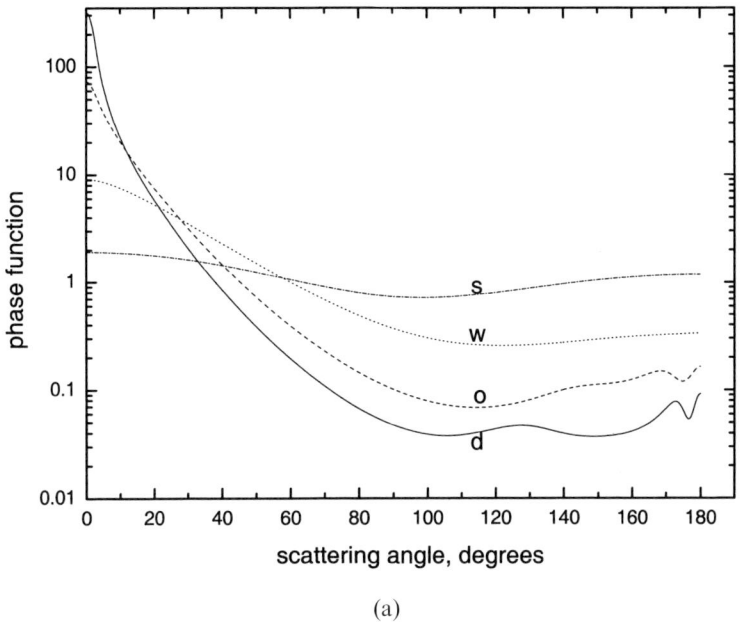

(a)

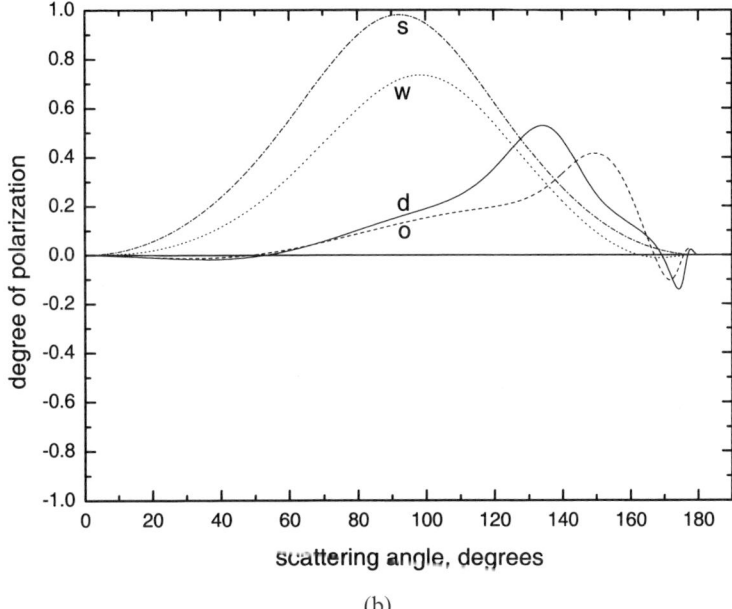

(b)

Figure 4.8. (a) The phase function of the water-soluble (w), soot (s), oceanic (o), and dust (d) aerosols at the wavelength 2.25 μm; (b) the same as in (a) except for the degree of polarization p_l.

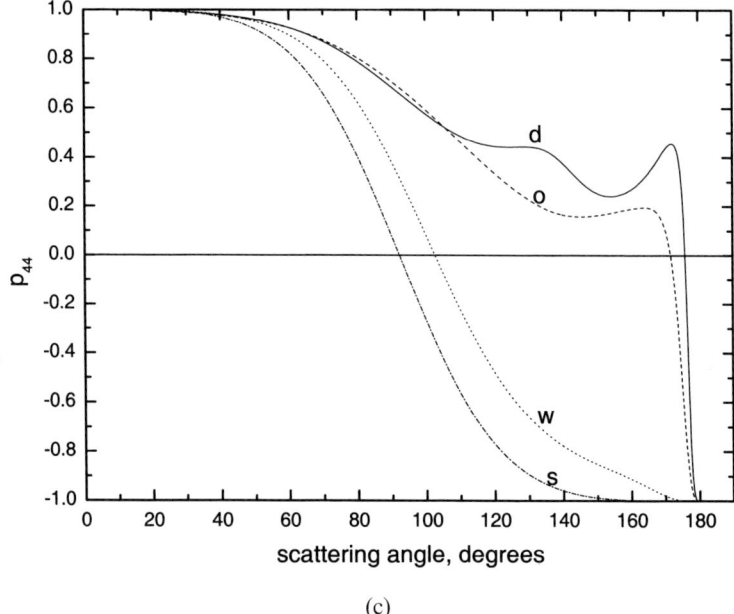

(c)

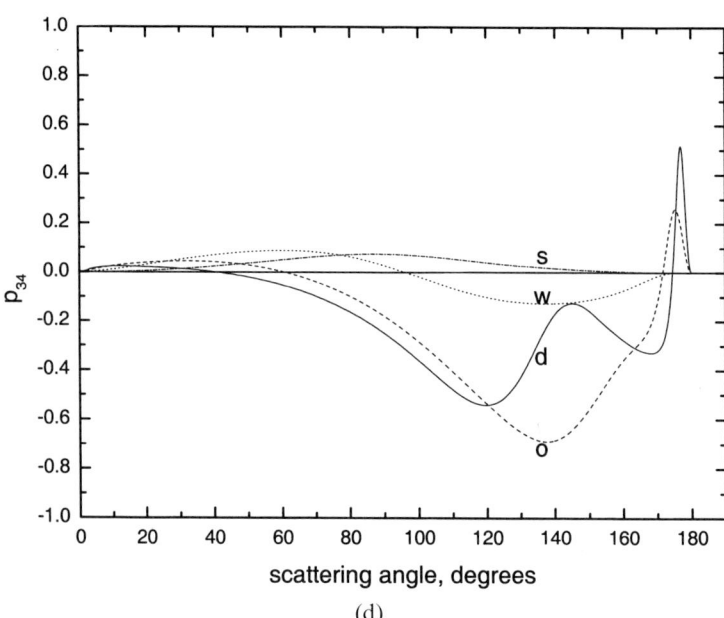

(d)

Figure 4.8 (*cont.*). (c) The same as in (a) except for p_{44}; (d) the same as in (a) except for p_{34}.

Sec. 4.2] Aerosols 165

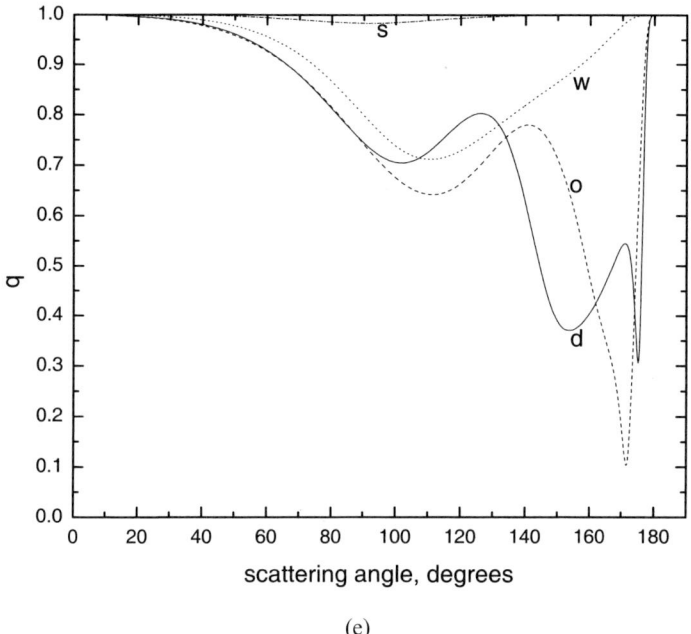

(e)

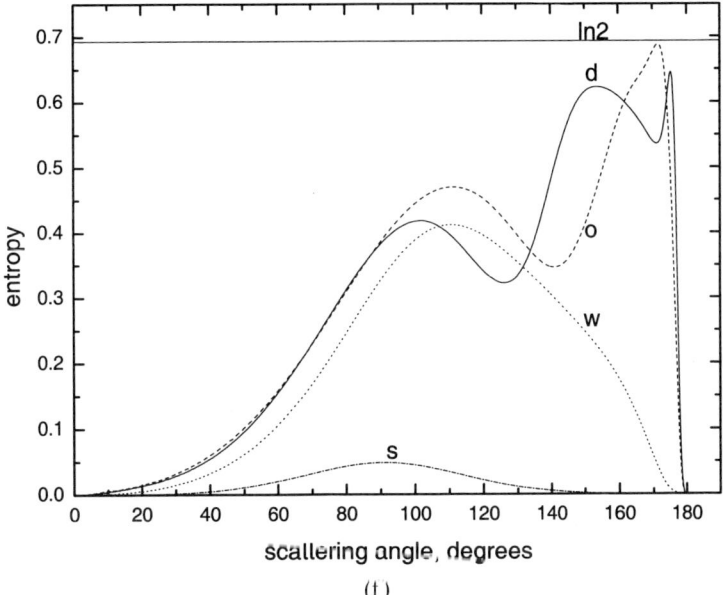

(f)

Figure 4.8 (*cont.*). (e) The same as in (a) except for q; (f) the same as in (a) except for the entropy s.

Table 4.2. Aerosol models (WCP-112, 1986).

Aerosol model	Volumetric concentration of the aerosol types given in Table 4.1
Continental	w – 29%, d – 70%, s – 1%
Maritime	w – 5% , o – 95%
Urban/industrial	w – 61%, d – 17%, s – 22%

Depending on the concentration of components in this mixture, various polarization features are more or less pronounced. For instance, the World Meteorological Organization model advises the use concentrations of various components in different areas of the globe as presented in Table 4.2.

It should be remembered that in practice, observed values of the degree of polarization for aerosol media can be different from those presented in this chapter due to the non-spherical shape of particles, influence of multiple light scattering, and ground reflection.

Let us consider the results of calculations in some detail. We start from the visible region of the electromagnetic spectrum. The phase function of soot aerosol is close to the Rayleigh phase function (see Figure 4.7(a)). Water-soluble aerosol is characterized by a smooth function in contrast to oceanic and dust aerosols, which have enhanced scattering towards the backward direction. It should be remembered, that in reality both dust and oceanic aerosols contain a large amount of non-spherical particles, which will make the phase functions more smooth in the backward hemisphere. We ignore these effects here, however.

The degree of polarization of soot aerosols is similar to the Rayleigh case, but with a somewhat lower value at the scattering angle of 90° (see Figure 4.7(b)). Also note the shift of the angle of the maximum polarization towards larger scattering angles. Oceanic aerosol is characterized by the negative (and small polarization) in the forward hemisphere, which is due to the directly transmitted component. Dust and water-soluble aerosols are characterized by quite large absorption coefficients. So directly transmitted rays make a small contribution. This is the reason for the positive polarization for dust aerosols in the forward scattering hemisphere. The maxima of polarization for dust and oceanic aerosol at large backscattering angles are due to rainbow scattering. The positions of rainbows are different due to the difference in the refractive index of particles. The dependence $p_{44}(\theta)$ for dust and oceanic aerosol behaves in a similar way, having a maximum near the scattering angle of 170°. The element p_{34}, which is responsible for the linear to circular light conversion, is rather small. It increases, however, for oceanic and dust aerosols in the backward hemisphere.

The degree of polarization q is close to unity for soot particles (see Figure 4.7(e)). However, it is reduced for oceanic, water-soluble, and dust aerosols in the backward hemisphere. The entropy of scattered light beams depends on the type of aerosols (see Figure 4.7(f)). It is higher for dust aerosols almost everywhere except for angles around 130°, where water-soluble aerosols produce large entropy of scattered light.

Similar results at the wavelength 2.25 µm are given in Figure 4.8. We see that the phase functions become smoother. The same applies to the degree of polarization. This is due to the decrease of the effective size parameter ka_{ef} of the droplets and larger absorption in the infrared as compared to the visible region of the spectrum. The element p_{44} for water-soluble aerosols approaches that for soot aerosols at this wavelength (see Figure 4.8(c)). This element changes sign only once for dust aerosols as compared to the visible range. Also the value of p_{44} is positive for oceanic and dust aerosols at all angles except for a small range near the backward direction (around 175–180°). This means in particular, that these aerosols at this wavelength (and outside the region of scattering angles close to 180°) do not change the sense of rotation of incident circular polarized light. Note that elements p_{34} look quite different for these wavelengths. In particular these elements are very small both for soot and water-soluble aerosols. They are larger for coarse aerosols. However, possible non-sphericity of the particles can reduce p_{34} both for oceanic and dust aerosols. This indicates that atmospheric aerosols are not effective producers of circular polarization even for incident linearly polarized beams. We see also that entropy production generally rises with the scattering angle (see Figure 4.8(d)). However, it is exactly zero for the forward and backward scattering directions.

Measurements of the value of q of atmospheric air can be used for the determination of the weather conditions. This was proposed by Gorchakov (1971). Figure 4.9 shows that the degree of polarization q is a very sensitive parameter to weather conditions. Gorchakov (1971) also established the correlation between the minimum of q and maximum of the degree of linear polarization p_l.

Note that to account for the molecular scattering contribution one needs to consider the phase matrix

$$\hat{P} = \frac{\hat{P}_A + \varsigma \hat{P}_R}{1 + \varsigma},$$

where $\hat{P}_A$ is the aerosol phase matrix studied above, $\hat{P}_R$ is the phase matrix of Rayleigh scattering, and ς is the ratio of the Rayleigh σ_{sca}^R and aerosol σ_{sca}^A scattering coefficients:

$$\varsigma = \frac{\sigma_{sca}^R}{\sigma_{sca}^A}.$$

We give the degree of polarization of light transmitted through Rayleigh atmosphere ($\sigma_{sca}^A = 0$) in Figure 4.10(a) for the incidence angle 53.1°. The degree of polarization decreases with the optical thickness. It takes a maximum at the angle equal to 90° counted from the solar angle. Also, we see points of zero polarization. These features are also present in results taken from measurements of the distribution of the degree of polarization of diffuse radiation transmitted by the terrestrial atmosphere. However, they are modified by the presence of the aerosols (Gorchakov, 1966) and ground reflection (Sekera, 1956). Both factors decrease the value of the degree of polarization. Therefore, curves given in Figure 4.10 should be considered as an possible upper limit for a given τ.

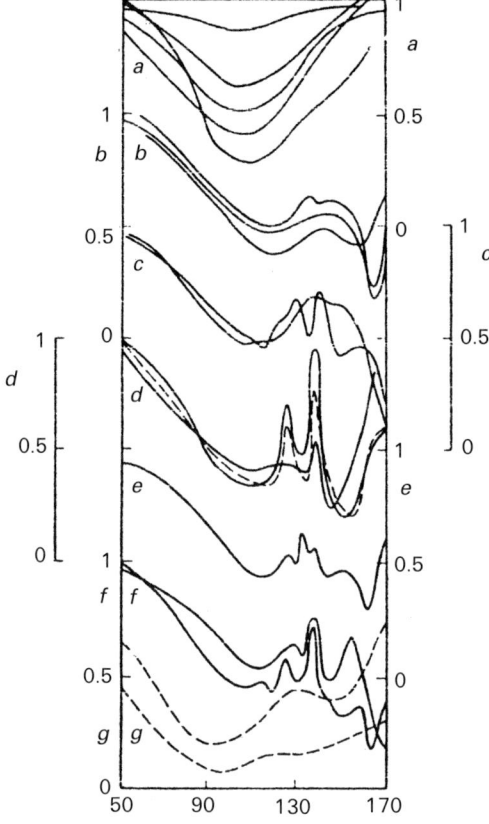

Figure 4.9. The dependence $q(\theta)$ according to measurements (Gorchakov, 1971) for different weather conditions at the wavelength 0.55 μm (a – haze, b – foggy haze, c – mist, d – haze with drizzle, e – foggy haze with drizzle, f – fog with drizzle, g – haze with snow).

Skylight polarization spectra at zenith direction were measured by Stam *et al.* (2002). They are shown in Figure 4.10(b) for the case of the diffusely transmitted light at solar angles equal to 68.3 and 50.4°. We see that the degree of polarization is in the range 0.23–0.3 for the smaller solar zenith angle. It is in the range 0.34–0.5 for the solar angles of 68.3°. Therefore, the degree of polarization increases with solar angle at the zenith observation. This is in full correspondence with Figure 3.10(a), which shows that the maximum of the degree of polarization should be when the difference between solar and observation angle is 90°. Note that the solar angles in Figure 4.10(a) and 4.10(b) (lower curve) almost coincide. The value of the degree of polarization should be equal to approximately 0.4 for the Rayleigh optical thickness 0.15 (see Figure 4.10(a)). However, it is smaller in reality (see Figure 4.10(b)). This is best explained by the influence of aerosols and ground reflection as explained above.

The spectra in Figure 4.10(b), which cover the wavelength region from 350 to

Sec. 4.2] Aerosols 169

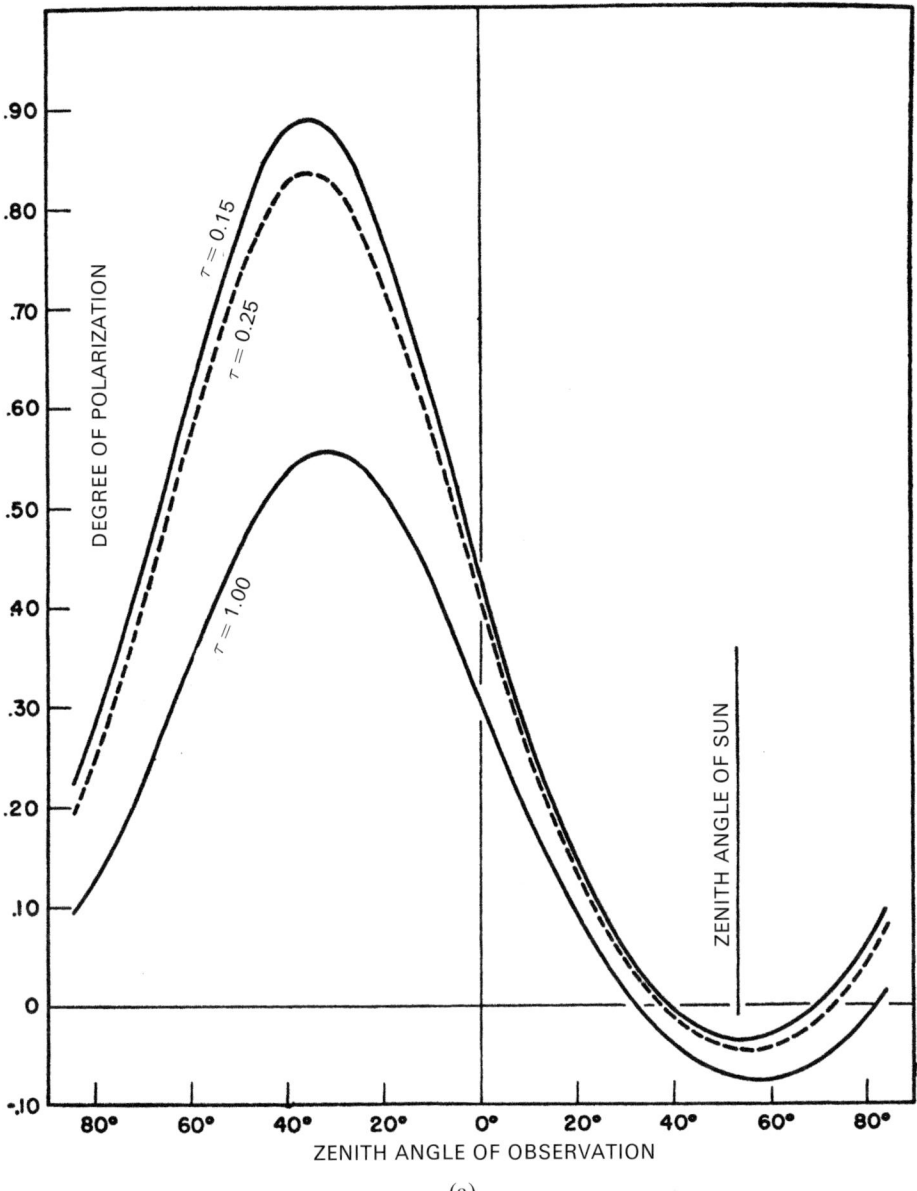

Figure 4.10. (a) Distribution of the degree of polarization along the sun's vertical axis in a Rayleigh atmosphere for a different optical thickness at the zenith distance of the sun of $53.1°$ (Sekera, 1956).

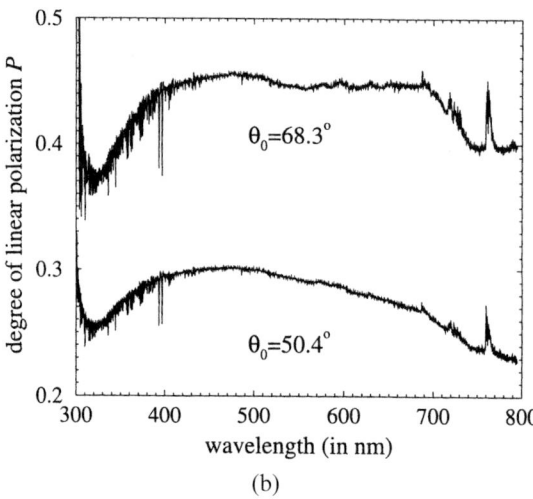

Figure 4.10 (*cont.*). (b) The degree of linear polarization of zenith skylight. For the upper curve, which was obtained on 3rd June 1997, the average solar zenith angle ϑ_0 is 68.3°; for the lower curve, obtained on 7th April, 1997, $\vartheta_0 = 50.4°$ (Stam et al., 2002).

400 nm, show a spectral fine structure that coincides with the Fraunhofer lines in the solar spectrum (Stam *et al.*, 2002). Observations show that the strength of the fine structure not only depends on the solar zenith angle but also on atmospheric conditions, in particular the aerosol content.

4.3 NATURAL SURFACES

Light reflected from natural surfaces like soil, snow, and vegetation is usually only weakly polarized. This is because such media can be considered as semi-infinite. As it is known, multiple scattering in optically thick media substantially diminishes polarization, which (for unpolarized illumination) is mostly due to the single scattering process. Clearly, for weakly absorbing surfaces, the role of multiple scattering is more important than for highly absorbing materials. Therefore, generally the polarization of reflected light for more absorbing media is somewhat larger.

It should be stressed that in contrast to atmospheric particulates, surfaces also have a specular reflection component (e.g., ocean surface). Therefore, one should consider not only light which emerges from inside the scattering medium in this case, but also the Fresnel reflection contribution. One can think that this is a minor problem because Fresnel reflection and refraction take place only in specific directions. In many practical situations this is not the case, however. Let us take a rough ocean under an overcast sky, for instance. Then the ocean is illuminated more less uniformly from all directions by almost unpolarized light. Therefore, specular reflection from the slopes of the waves can give quite a large value of light polarization for specific geometries.

Characteristics of light polarization are different for downward and upward propagation directions (with respect to the light source). For instance, a non-zero ellipticity e of light propagated in the downward direction in the ocean was detected (Ivanoff and Waterman, 1958a, b). It rapidly diminishes with the depth and almost equals zero in reflected light. The value of $e \neq 0$ is mainly due to total reflection at the water surface of linearly polarized light arising from the scattering of directional light in the water. This feature is used by oceanic creatures for navigation purposes.

At large depths in the ocean the polarization is quite small and is characterized by an asymptotic distribution, which does not depend on the depth. The polarization pattern is symmetric with respect to the vertical. The light is partially linearly polarized in this case. It is believed that a small portion of light is elliptically polarized at the large depth due to optical activity of oceanic bio-suspensions. However, this has not been confirmed by experiments so far.

4.3.1 Land

For remote sensing applications, the land is usually considered as a linear mixture of bare soil and vegetation. Then the reflection function R is given approximately as a combination of reflection from soil R_s and vegetation R_v:

$$R = cR_s + (1-c)R_v,$$

where c can be determined using so-called normalized differential vegetation index. Spectra $R_s(\lambda)$ and $R_v(\lambda)$ differ considerably (see Figure 4.11). The same applies to polarization curves for both media. Note, for instance, the sharp increase of the reflection for vegetation around 700 nm.

The degree of polarization for various mineral materials is given in Figure 4.12. The Yolo loam sample is a fine-grained loam soil from Yolo County, California. Red clay is the subsurface soil of Pennsylvania. The limestone and limonite samples were pulverized from the bulk minerals. Note that the degree of polarization generally depends on the degree of pulverization. Smaller grains are generally weaker absorbers. Therefore, they reduce the degree of polarization.

It can be seen in Figure 4.12 that all the curves are similar. The polarization maximum is located in the range 95–110° from the anti-source (see the arrow in Figure 4.12) and the polarization is in the range 7–20% in this case. Note also the small (and negative) polarization in the vicinity of the anti-source direction.

The degree of polarization of light reflected from desert sand is given in Figure 4.13 for a range of wavelengths. The curves are similar to those given in Figure 4.12 but the polarization is somewhat larger, which is most probably due to large absorption in sand as compared to the media in Figure 4.12. It should be pointed out that the degree of polarization increases with the soil water content. The reflection function is also affected as can be easily observed for dry and wet sand (e.g., at the beach). The degree of polarization of light reflected from black loam soil is given in Figure 4.14. A strong polarization at the shorter wavelengths is evident.

The polarization characteristics of land covered by vegetation differ from those for a bare soil. The spectral dependence of the degree of polarization for light

172 Environmental polarimetry   [Ch. 4

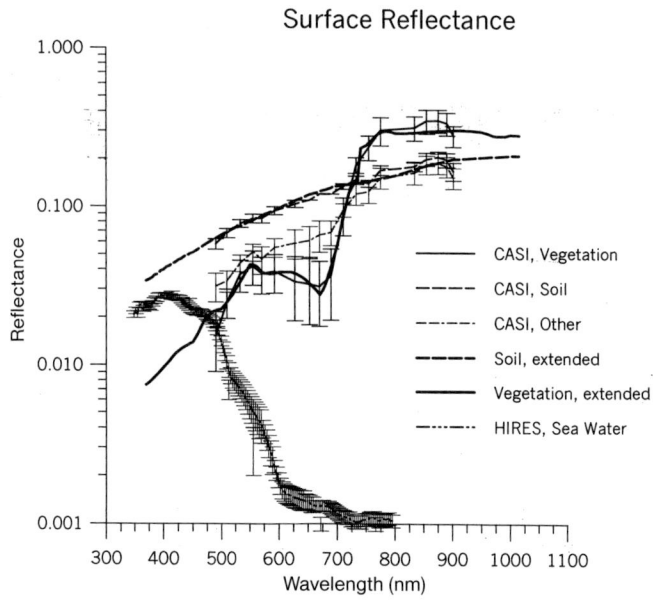

Figure 4.11. Typical reflectance spectra of land and water surfaces, composed from CASI-radiometer measurements and extensions, taken from the CAMELEO database (von Hoyningen-Huene et al., 2003). For water surfaces of the Baltic Sea, data from measurements of the HIRES radiometer (Institute of Space Techniques, DLR) are presented. The reflection of light by water at $\lambda > 700$ nm can be neglected.

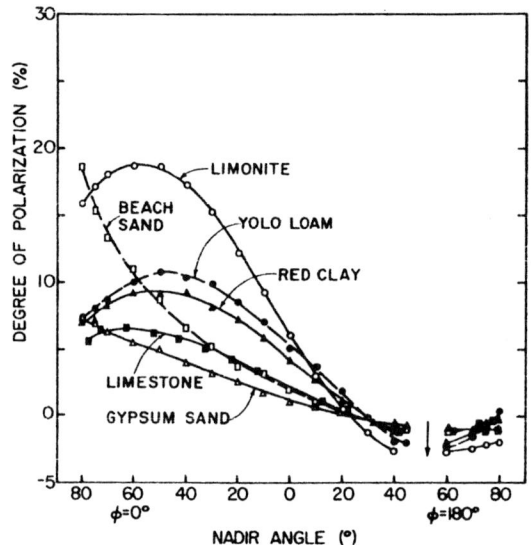

Figure 4.12. Degree of polarization as a function of angle in the principle plane of light reflected from various types of mineral surfaces for the angle of incidence 53° at the wavelength 0.492 μm (Coulson, 1966).

Sec. 4.3]   Natural surfaces 173

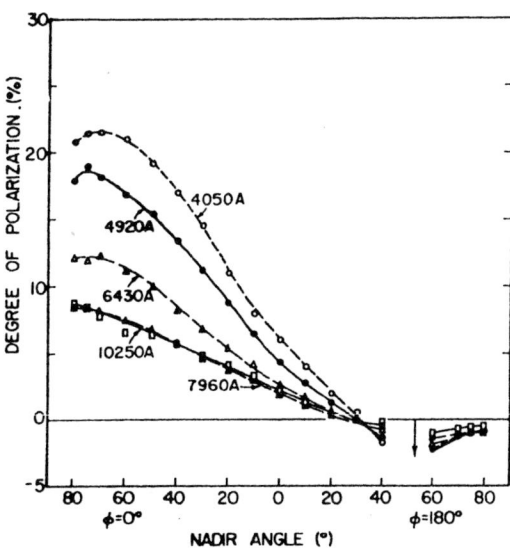

Figure 4.13. Degree of polarization as a function of the angle in the principal plane for light of five different wavelengths reflected from the Mojave Desert sand for the angle of incidence 53° (Coulson, 1966).

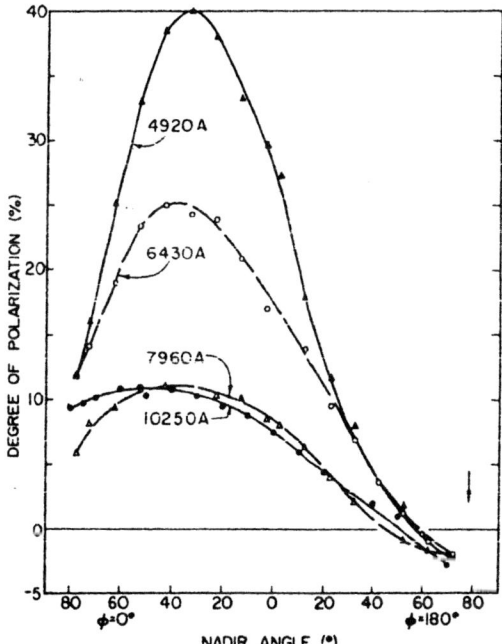

Figure 4.14. Degree of polarization as a function of angle in the principal plane of light of four different wavelengths reflected from black loam soil for the angle of incidence of 78.5° at the wavelength 0.492 µm (Coulson, 1966).

reflected from green, red, and white portions of leaves differs. This brings an additional complexity to the problem. Otherwise, this can be used for plant conditions remote monitoring (e.g., for the indication of diseases, excessive salinity). Note that specular reflection of light from the leaf is also an important source of light polarization.

4.3.2 Ocean

Most of the surface of our planet is covered by ocean. Light polarization characteristics of ocean water have been studied by many authors. Ocean can be modelled as a semi-infinite vertically stratified medium with horizontally variable properties. We will neglect the influence of horizontal and vertical inhomogeneity and assume that the water–air interface is smooth. This is, of course, an idealization of a real situation, which could involve breaking waves, whitecaps, and schools of fish. However, we are interested in understanding some basic features. Therefore, we ignore all possible complications. Still, the problem is not very easy to handle. It requires the solution of a system of four integro-differential radiative transfer equations. For this we need to specify the reflection matrix of an interface, which polarizes incident and escaping light. This is done quite easily using the Fresnel equations (Mobley, 1994).

The problem of phase matrices for oceanic water is more involved. These can be presented as combinations of the phase matrices for molecular scattering in water and scattering by various particles in the water (e.g., plankton, dendrites, and bubbles). This varies across the ocean.

The model of spherical particles which is used even in up-to-date studies in ocean optics (Stramski et al., 2001) cannot be applied if one has a goal to describe the phase matrix (Bohren and Huffman, 1983) of oceanic water. For instance, the element P_{22} of this matrix is equal to P_{11} for spherical polydispersions. This is not confirmed by measurements of both open ocean and coastal areas (Kadyshevich et al., 1974; Voss and Fry, 1984).

Clearly, the theory of light scattering by non-spherical particles should be used in this case. However, both the complexity of the theory and the high variability of ocean water microstructure makes the solution of the problem highly complicated. With this in mind, we propose here the parameterization of the phase matrix of oceanic waters, based on experimental measurements.

The normalized phase matrix of oceanic water has the following general form (Bohren and Huffman, 1983):

$$\hat{p} = \begin{pmatrix} 1 & p_{12} & p_{13} & p_{14} \\ p_{21} & p_{22} & p_{23} & p_{24} \\ p_{31} & p_{32} & p_{33} & p_{34} \\ p_{41} & p_{42} & p_{43} & p_{44} \end{pmatrix},$$

where elements $p_{13}, p_{14}, p_{23}, p_{24}, p_{31}, p_{32}, p_{41}, p_{42}, p_{34}$, and p_{43} take very small values and can be neglected for most studies, which, of course, simplifies the theoretical analysis. However, they should not be considered as exact zeros in general. They could be of importance in the solution of selected problems in ocean optics, especially those focused on chiroptical spectroscopy. Also, there are experimental indications that these elements can have quite large values in individual cases (Kadyshevich et al., 1974). This could be due to the orientation and chirality of the particles. However, we will neglect all these effects and consider the matrix to be of the form:

$$\hat{p} = \begin{pmatrix} 1 & p_{12} & 0 & 0 \\ p_{21} & p_{22} & 0 & 0 \\ 0 & 0 & p_{33} & 0 \\ 0 & 0 & 0 & p_{44} \end{pmatrix}.$$

This is confirmed by the experiments of Voss and Fry (1984). They also found that $p_{12} \approx p_{21}$ and $p_{33} \approx p_{44}$. Then the normalized phase matrix takes the following simple form

$$\hat{p} = \begin{pmatrix} 1 & c & 0 & 0 \\ c & a & 0 & 0 \\ 0 & 0 & b & 0 \\ 0 & 0 & 0 & b \end{pmatrix}.$$

We are interested in modelling the dependencies of the elements a, b, and c on the scattering angle. They have the following analytical descriptions for the idealized case of oceanic waters free of impurities (Shifrin, 1988):

$$a_m = -\frac{c_m\left(\frac{\pi}{2}\right)(1 + \cos^2\theta)}{1 - c_m\left(\frac{\pi}{2}\right)\cos^2\theta},$$

$$b_m = -\frac{2c_m\left(\frac{\pi}{2}\right)\cos\theta}{1 - c_m\left(\frac{\pi}{2}\right)\cos^2\theta},$$

$$c_m = \frac{c_m\left(\frac{\pi}{2}\right)\sin^2\theta}{1 - c_m\left(\frac{\pi}{2}\right)\cos^2\theta},$$

where the subscript 'm' underlines the fact that these formulae describe only molecular light scattering. $c_m(\pi/2) = -0.835$ at a wavelength of 436 nm and a temperature of 20°C (Shifrin, 1988).

Accounting for particle scattering modifies the equations given above. In particular, values of a, b and c can be presented in the following form in this case:

$$a = \frac{-c\left(\frac{\pi}{2}\right)(1+\cos^2(\theta-\varsigma\theta_0))+\varsigma p(\theta)}{1-c\left(\frac{\pi}{2}\right)\cos^2(\theta-\varsigma\theta_0)+\varsigma p(\theta)},$$

$$b = \frac{-2c_m\left(\frac{\pi}{2}\right)\cos\theta+\varsigma p(\theta)}{1-c_m\left(\frac{\pi}{2}\right)\cos^2\theta+\varsigma p(\theta)},$$

$$c = \frac{c\left(\frac{\pi}{2}\right)\sin^2\theta}{1-c\left(\frac{\pi}{2}\right)\cos^2\theta}.$$

Here $\varsigma\theta_0$ is the shift of the minimum in the function $a(\theta)$, which is due to particle scattering. The parameter $\varsigma\theta_0$ has a clear physical sense and can be easily measured.

Note that $p_l(\pi/2) = -c(\pi/2)$ gives us the degree of polarization of scattered light at $\theta = \pi/2$ for the case of natural light illumination. We have as $\varsigma \to 0$: $a \to a_m$, $b \to b_m$, and $c \to c_m$ as it should be.

The equations given above can be written in the following form:

$$a(\theta) = \frac{p_l\left(\frac{\pi}{2}\right)(1+\cos^2(\theta-\varsigma\theta_0))+\varsigma p(\theta)}{1+p_l\left(\frac{\pi}{2}\right)\cos^2(\theta-\varsigma\theta_0)+\varsigma p(\theta)},$$

$$b(\theta) = \frac{2p_l\left(\frac{\pi}{2}\right)\cos\theta+\varsigma p(\theta)}{1+p_l\left(\frac{\pi}{2}\right)\cos^2\theta+\varsigma p(\theta)},$$

$$p_l(\theta) = \frac{p_l\left(\frac{\pi}{2}\right)\sin^2\theta}{1+p_l\left(\frac{\pi}{2}\right)\cos^2\theta}.$$

Comparisons of calculations using these formulae with the experimental data of Voss and Fry (1984) are presented in Figures 4.15(a)–(c), where the following representation of the function $p(\theta)$ was used:

$$p(\theta) = 2\alpha^2 \exp(-\alpha\theta).$$

Sec. 4.3] Natural surfaces 177

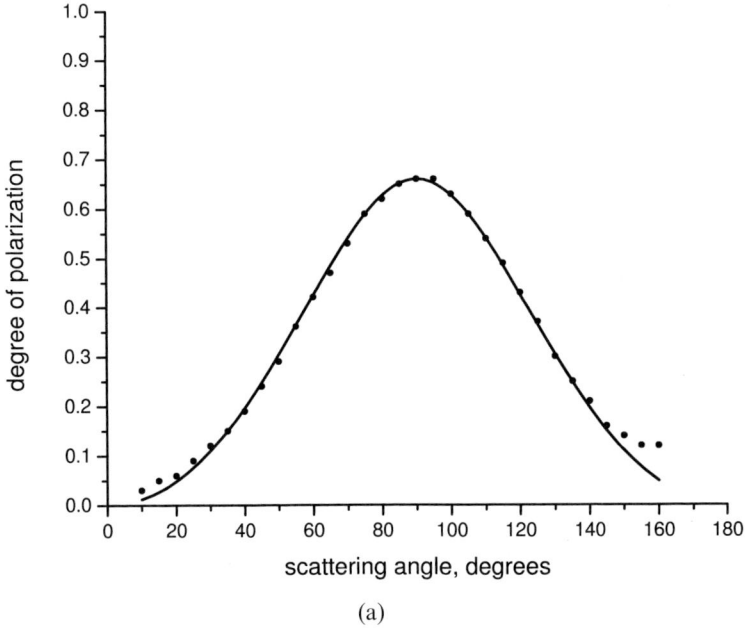

(a)

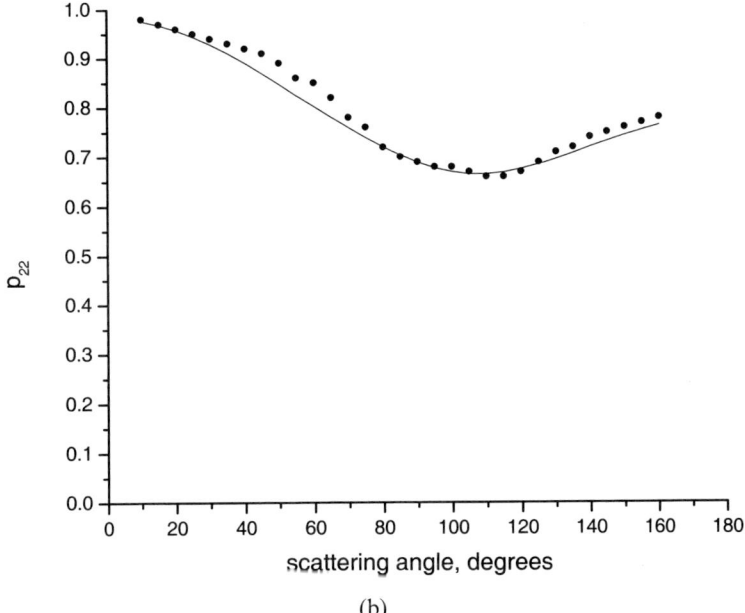

(b)

Figure 4.15. (a) The Voss and Fry (1984) data (points) for the degree of polarization of light scattered by an elementary volume of oceanic water. Results of our model are given by a solid line; (b) the same as in (a) except for p_{22}.

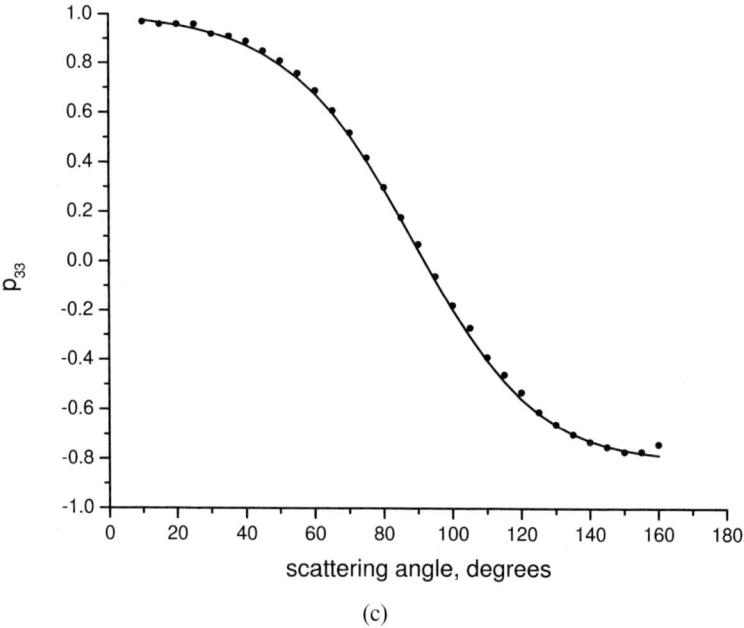

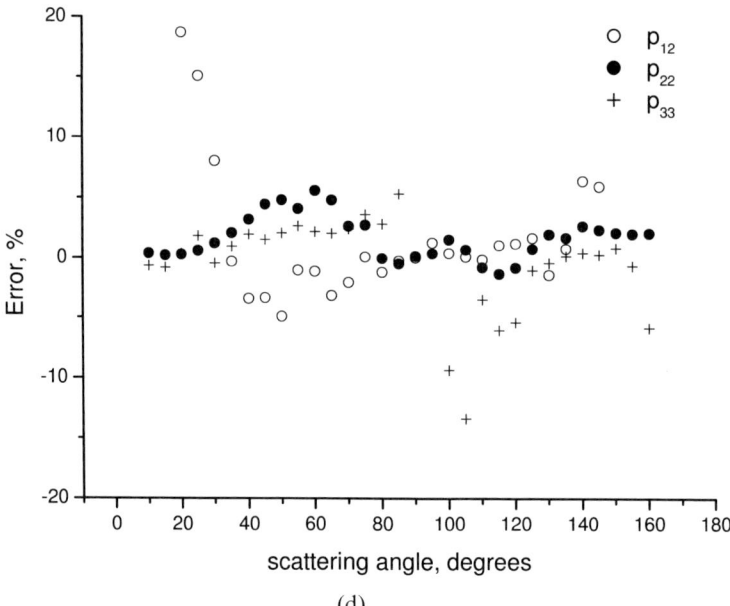

Figure 4.15 (*cont.*). (c) The same as in (a) except for p_{33}; (d) errors of approximate formulae, calculated from data given in (a)–(c).

The free parameters α, $p_l(\pi/2)$, θ_0 and ς were obtained by fitting experimental data. They appeared to have the following values:

$$\alpha = 4\,\text{rad}^{-1}, \qquad p_l(\pi/2) = 0.66, \qquad \theta_0 = \tfrac{1}{3}\,\text{rad}, \qquad \varsigma = \tfrac{4}{5}.$$

Note that the value of $p_l(\pi/2)$ correlates with the ratio of the volume scattering coefficient of particles to that of a pure water at $\theta = \pi/2$.

The relative errors of the approximations given here are generally smaller or comparable to the errors of the experiments involved (see Figure 4.15(d)). This confirms the fact that they capture correctly the behaviour of the functions $a(\theta), b(\theta), c(\theta) = -p_l(\theta)$ in natural oceanic waters.

Let us also underline the disagreement between the parameterization and measurements for $p_l(\theta)$ at small and large values of the scattering angle θ. However, the degree of polarization here is extremely low, which makes possible the influence of numerous factors that are out of reach of the parameterization considered here. Large errors in $b(\theta)$ at $\theta \sim \pi/2$ are of no importance. This is just due to the fact that $b(\pi/2) = 0$. Thus the absolute errors and not relative ones should be considered in this particular case.

Specular reflection of light from the ocean can be studied with the Fresnel equations given in Appendix 5. Note that the specular reflection of pure and contaminated (e.g., by oil ($n \approx 1.57$)) water is different. This is used for oil film detection on oceanic surface.

4.3.3 Snow and ice

Ice clouds can precipitate, forming a layer of ice crystals on the surface of our planet. Initially particles in snow coincide in shape and structure with the ice particles from

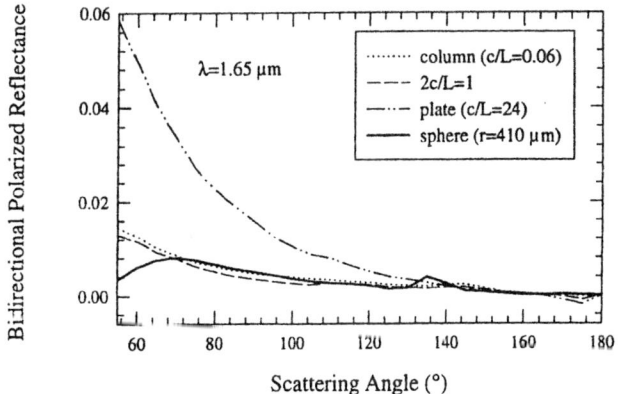

Figure 4.16. The bidirectional polarized reflectance (the Stokes parameter Q) as the function of the scattering angle for semi-infinite snow surface at the wavelength 1.65 μm. Three shapes of particles were considered, namely: ice columns, plates, and spheres (Leroux et al., 1999). The solar angle is equal to $40°$.

precipitating clouds. They change with time, producing a random medium quite different from ice clouds. First of all, particles in snow are much closer to each other than in the case of ice clouds. Also the grains grow with time. It was found that the largest grains are closest to the surface. They easily accumulate liquid water on warm days. Various aerosols (especially in industrial areas) penetrate the snow layer, increasing its absorbing power.

Results of calculations of the bidirectional polarized reflectance Q of snow are given in Figure 4.18 for particles of various shapes at $\lambda = 1.65\,\mu\mathrm{m}$. The value of Q is almost constant and close to zero for all angles in the backward hemisphere. This suggests that snow does not produce considerable amounts of polarized light in the visible and near-infrared regions of the electromagnetic spectrum.

This is not the case for snow layers for larger wavelengths, where multiple light scattering is weak. Also, note that snow can be covered by a thin ice layer. Then the polarization of reflected light increases.

It should be emphasized that the theoretical treatment of polarized light transfer in natural surfaces has a lot of important obstacles. One of them is the high density of particles in such media. For instance, the volumetric concentration of snow grains is around 30%. This puts forward the question as to what extent the radiative transfer equation can be used in such applications. Certainly, the extinction and scattering coefficients and the phase matrix generally depend on the spatial correlation of the scatterers. This is not accounted for in the standard radiative theory.

4.4 POLARIMETRIC REMOTE SENSING

Studies of the polarization characteristics of solar light transmitted and reflected by aerosols and cloudy media have a long history. However, the real burst of research in this area was given by the launch of the POLDER (Polarization and Directionality of Earth Reflectances) instrument on board the Japanese *ADEOS-I* and *ADEOS-II* satellites. The POLDER was able to transmit to the Earth a huge amount of information about polarization characteristics of light reflected from cloudy media, aerosols, and underlying surfaces at several wavelengths. Specifically, the first three components of the Stokes vector $\vec{S}_r(I, Q, U, V)$ were measured for wavelengths λ equal to 443, 670, and 865 nm. There is no doubt that even more advanced polarimeters with more wide spectral coverage will appear on board other satellites in future, which makes further theoretical studies of polarization characteristics of natural surfaces, clouds, and aerosols important. This is due to the potential possibilities for global retrieval (e.g., of cloud and aerosol microstructure, the shape of particles and the optical thickness of clouds based on polarization measurements (Masuda *et al.*, 2002a, b)).

Of course, similar information can be obtained from reflected light intensity measurements. However, it could well be that the degree of polarization can be used as a source of additional information about particle size distributions close to the top of a cloud. Multiply scattered light fluxes from deep layers are hardly polarized at all. Radiative characteristics, on the other hand, represent the cloud as a whole. Therefore, the effective radius derived from radioactive measurements corresponds to an average radius obtained for large ensembles of possibly very different particle size distributions.

It has been emphasized by Hansen (1971) that the fact that the polarization data refer to cloud tops may not be a drawback but an advantage. Indeed, it is easier to interpret observations which do not refer to average values over the entire cloud depth.

The polarization characteristics of aerosol and cloudy media can be studied by applying numerical codes based on the vector radiative transfer equation solution. However, one can also use the fact that cloud fields are optically thick in most cases. This allows asymptotic analytical relations, derived for optically thick disperse media with arbitrary phase functions and absorption (see Chapter 2) to be applied. These solutions help us to clarify the physical mechanisms and main features behind the polarization change due to the increase of the size of droplets or the thickness of cloud. Analytical solutions also provide an important tool for the simplification of the inverse problem. They can be used, for example in studies of the information content of polarimetric measurements.

The drawback of the well known vector asymptotic theory for optically thick layers is related to the fact that the main equations depend on different auxiliary functions and parameters. Calculations of these functions and parameters are rather complex. This is due to their quite general applicability in terms of the single scattering albedo.

The special case of weakly absorbing light scattering media was considered in the framework of the modified asymptotic theory in Chapter 2. The formulae obtained are rather simple and allow for a semi-analytical solution of the inverse problem. In particular, the equations obtained can be used for a semi-analytical retrieval of droplet sizes from the two-wavelength polarimetric inversion algorithm (Kokhanovsky and Weichert, 2002). Let us rewrite these equations here, explicitly showing the wavelength dependence:

$$P(\lambda_1) = \frac{P^0_\infty(\lambda_1)}{1 - ut(\lambda_1)},$$

$$P(\lambda_2) = \frac{P^0_\infty(\lambda_2)\exp(s(\lambda_2)y(\lambda_2))}{1 - ut(\lambda_2)\exp(-x(\lambda_2) - y(\lambda_2)(1 - s(\lambda_2)))},$$

where λ_1 is the wavelength in the visible λ_2 is the wavelength in the infrared. A closer look at these equations shows that values $P(\lambda_1)$ and $P(\lambda_2)$ depend only on the solar elevation, optical thickness at each wavelength, the droplet size distribution, and the refractive index. It is known, however, that the influence of the type of the particle size distribution on the integral scattering characteristics, such as x and y, is rather weak. So we will neglect this dependence and characterize the size of droplets by just one number – the effective radius of droplets. We also ignore the possible variability of the halfwidth of the droplet size distribution and use the distribution (1.5) at $\mu = 6$

in the discussion which follows. The dependence of the degree of polarization of the singly scattered light on the value of μ is presented in Chapter 3. We see that this dependence can be neglected for most scattering angles.

The functions $P_\infty^0(\lambda_1)$, and $P_\infty^0(\lambda_2)$ at wavelengths $\lambda_1 = 0.65\,\mu m$ and $\lambda_2 = 1.55\,\mu m$ depend on the PSD only. They can be parameterized in terms of the cloud droplet effective size as follows:

$$P_\infty^0(\lambda) = A - \frac{B}{1 + \exp(\sigma_1 + \sigma_2 a_{ef})}.$$

The parameters A, B, σ_1, and σ_2 depend on the wavelength. In particular, at $\lambda = 0.65\,\mu m$: $A = 9.1$, $B = 22$, $\sigma_1 = 0.37$, and $\sigma_2 = 0.17$. Also at $\lambda = 1.55\,\mu m$: $A = 5.5$, $B = 8.4$, $\sigma_1 = -1.0$, and $\sigma_2 = 0.29$. The accuracy of this parameterization is better than 6% for effective radii in the range 4–16 μm (Kokhanovsky and Weichert, 2002).

These two wavelengths are almost free of gaseous absorption and can be used in the development of the inversion algorithm. Thus, we have the system of two equations given here to find the optical thickness in the visible and the size of the droplets a_{ef}. We proceed as follows. First of all we express the value $\tau(\lambda_1)$ via the measured degree of polarization $P(\lambda_1)$:

$$\tau(\lambda_1) = \frac{4[t^{-1}(\lambda_1) - \alpha]}{3[1 - g(\lambda_1)]},$$

where

$$t(\lambda_1) = \frac{1}{u}\left\{1 - \frac{P_\infty^0(\lambda_1)}{P(\lambda_1)}\right\}.$$

Secondly, we use the fact that the ratio $\Upsilon(a_{ef}) = \tau(\lambda_2)/\tau(\lambda_1)$ depends only on the effective radius of droplets (see Appendix 4). So we can substitute the value of $\tau(\lambda_2)$ by $\Upsilon(a_{ef})\tau(\lambda_1)$. This gives us a single transcendental equation for the effective radius determination, which can be easily solved using standard techniques (Kokhanovsky and Weichert, 2002). The value of $\tau(\lambda_1)$ can be obtained afterwards. This completes the proposed inversion algorithm description.

Let us study the errors associated with this retrieval algorithm. We will use the fixed viewing geometry ($\vartheta_0 = 37°$, and $\vartheta = 0°$), where the sensitivity of the measured degree of polarization in the visible to the value of the effective radius of droplets is rather high. However, one could also choose another geometry. This is mostly due to the fact that the information on the droplet sizes is contained in the parameter y, which does not depend on the observation geometry. This is an important point.

We have calculated the values of $P(\lambda_1)$ and $P(\lambda_2)$ at $a_{ef} = 6\,\mu m$ and $\tau = 10$ with the exact radiative transfer code for wavelengths 0.65 and 1.55 μm. Then we assumed that the actual values of $P(\lambda_1)$ and $P(\lambda_2)$ differ from those calculated by the relative positive error Δ, which was varied from 0 to 10%. The errors of retrieval obtained are presented in Figure 4.17. We see that the errors in the retrieved values

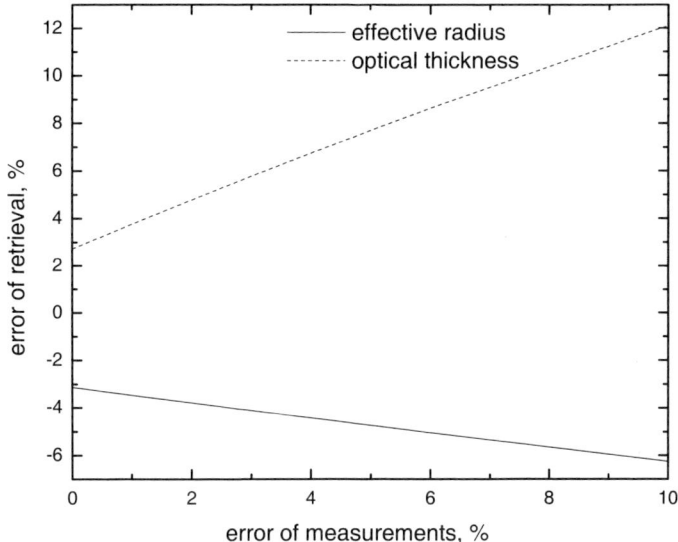

Figure 4.17. The error of retrieval of values of effective radius and cloud optical thickness as a function of the measurement error.

of a_{ef} and τ are smaller than 12% in this case. This shows that our algorithm is not particularly sensitive to the measurement errors at $a_{ef} = 6\,\mu\text{m}$ and $\tau = 10$. Clearly, the errors of retrieval of the optical thickness increase with the optical thickness. This is due to the fact that the degree of polarization for extremely thick clouds does not dependent on the optical thickness at all. So it cannot be retrieved in principle. This is not the case as far as the effective radius of the droplets is concerned. This can be derived for clouds of arbitrarily large thickness. However, for large clouds the transmission t is approximately equal to zero and the transcendental equation that needs to be solved transforms into a more simple form which can be solved analytically.

Numerically solving the single transcendental equation specified above, one obtains the value of a_{ef}, which subsequently can be used to find the optical thickness and liquid water path of clouds (Kokhanovsky and Weichert, 2002) if they are not very thick. Clouds with optical thicknesses around 15 and larger could be considered as semi-infinite for wavelengths larger than $1.55\,\mu\text{m}$. Of course, this is not the case in the visible range of the electromagnetic spectrum. Knowing values of a_{ef} and τ, it is easy to obtain many other parameters of cloudy media, including their transmittance, reflection function, and spherical albedo.

Spectral measurements at several wavelengths can improve the accuracy of the algorithm. Remarkably, however, there is even the possibility of finding the size of particles and liquid water path of clouds from measurements of the reflection function and the degree of polarization at a single wavelength in the visible spectrum. For this one should make measurements at the rainbow geometry (see Figure 4.18).

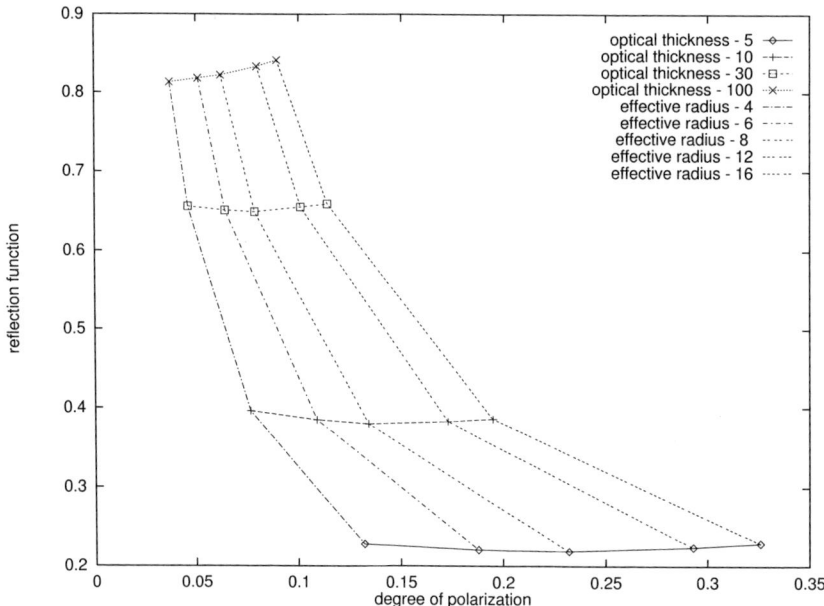

Figure 4.18. The dependence of the reflection function on the degree of polarization at the rainbow scattering angle for various cloud optical thickness and effective radii of droplets. The wavelength is equal to 0.67 μm. The observation is in the direction perpendicular to the cloud at the solar zenith angle to 37°.

Similar techniques can be applied to polarimetric sensing of aerosols (Zhang and Gordon, 1997; Zhao et al., 1997; Masuda et al., 1998, 2002a) and optically thin crystalline clouds (Masuda et al., 2002b). However, the asymptotic theory is not valid in this case and solutions of the vector radiative transfer equation should be used. Another problem is related to the fact that the non-spherical shape of particles should be fully taken into account (Mishchenko et al., 1996) and this is not an easy task, taking into account the complexity of shapes in ice clouds and dust aerosols.

4.5 CHIRO-OPTICAL SPECTROSCOPY OF TURBID MEDIA

4.5.1 Introduction

Ocean water and the leaves of various plants have living cells as inclusions which are characterized by intrinsic asymmetry. This leads to a rotation of plane of polarization of linearly polarized light incident propagating through such media. These effects are rather weak. However, they are detectable and can be used as important sources of information on living cells (e.g., the arrangement of chemical groups inside molecules during their life cycles).

Circular dichroism (CD) $\varphi(\lambda)$ and optical rotatory dispersion (ORD) $\psi(\lambda)$

spectra are sources of information on the asymmetry of molecular structures. The CD and ORD spectra are used in organic chemistry as a routine tool of stereochemical analysis.

In most heterogeneous cases chiro-optical spectroscopy (CS) is applied to substances in solution. However, biological media are usually highly microscopic due to the presence of various microscopic particles (e.g., red blood cells). Thus, there is a task of generalizing the CS to the case of particulate media.

The major result is that CD and ORD spectra of substances inside small particles may differ substantially from the spectra measured for dispersed layers. They become dependent on the refractive indices of particles, their size, shape and possibly, on the concentration of particles if the optical thickness of a sample is high. The case of chiral spheres can be treated in the framework of exact electromagnetic scattering theory as discussed in Chapter 3. However, most biological particles are non-spherical in shape. So, the generalization should be made allowing for the non-spherical shapes of the particles in an optically active turbid layer. This can be achieved in two ways. The first is to use known solutions of electromagnetic scattering problems for chiral cylinders and spheroids. The second is to apply approximate scattering theories (e.g., the Rayleigh, Rayleigh–Gans, discrete dipoles, and anomalous diffraction approximations (Kokhanovsky, 2001)). Here we will use the second approach, namely the anomalous diffraction theory (ADT) of Van de Hulst (1981).

This is mostly due to the following reasons. First, the ADT is perfectly suitable for the solution of applied problems due to its intrinsic mathematical simplicity. Second, the range of shapes which can be considered is wide. This is not the case for exact electromagnetic scattering solutions. In fact, solutions of the wave equation for most particle shapes occurring in nature have not been derived due to the mathematical complexity involved.

For instance, it was found that an additional multiplier is needed to account for light scattering effects in CS of disperse media with particles which are much smaller than the wavelength λ of the incident light (Bohren and Huffman, 1983). The case of arbitrarily shaped particles at $kd|\bar{m} - 1| \ll 1$, $|\bar{m} - 1| \ll 1$ (d is the maximal size of particles, $k = 2\pi/\lambda$, $\bar{m} = (m_L + m_R)/2$, m_L and m_R are refractive indices for left-handed and right-handed circularly polarized waves correspondingly) was studied by Bohren (1978) in the framework of the Born approximation. It was found that light scattering effects can be neglected in this particular case. Correspondingly, CD and ORD spectra of substances inside such particles coincide with spectra measured for dispersed layers (in the framework of the single scattering approximation, of course). This allows us to use the CS developed for homogeneous media in the case of particulate media

Clearly, the CD and ORD spectra of turbid layers are not equal to the CD and ORD spectra of substances inside particles if the condition $kd|\bar{m} - 1| \ll 1$ is not satisfied (either due to the small value of the wavelength λ or large values of d). Let us consider this discrepancy for the case of large ($kd \gg 1$) optically soft ($|\bar{m} - 1| \ll 1$) non-spherical particles, where the Van de Hulst (or anomalous diffraction) approximation can be used.

4.5.2 CD and ORD spectra of turbid layers with non-spherical particles

Let us assume that the optical thickness of the medium in question is low, which allows us to neglect the contribution of multiply scattered photons. Then the disperse medium can be considered (at least as far as the light transmission is concerned) as homogeneous with the effective refractive index (see Section 3.1.2):

$$M = 1 - 2\pi i N k^{-3} S(0), \quad (4.1)$$

where N is a number of particles in a unit volume and $S(0)$ is the so-called amplitude scattering function in the forward direction. Clearly, this equation can be applied only at small values of $|M - 1|$.

Equation 4.1 is easily generalized for the case of disperse media with optically active particles. Two refractive indices are introduced in this case: M_1 – for left-handed circularly polarized waves and M_2 – for right-handed circularly polarized waves. The values of M_1 and M_2 coincide for isotropic non-chiral particles. They generally differ for optically active particles, but the difference is usually small ($|M_1 - M_2| \ll 1$). Nevertheless, it is detectable and can be used to study molecular asymmetries of substances inside small light scattering globules. Thus:

$$M_j(\lambda) = 1 - 2\pi i N k^{-3} S_j(\lambda) \quad (4.2)$$

and by definition

$$\psi(\lambda) = \frac{\pi}{\lambda} \operatorname{Re}(M_1(\lambda) - M_2(\lambda)), \quad (4.3)$$

$$\varphi(\lambda) = \frac{\pi}{\lambda} \operatorname{Im}(M_1(\lambda) - M_2(\lambda)), \quad (4.4)$$

where $\psi(\lambda)$ and $\varphi(\lambda)$ are the ORD and CD spectra per unit length of a sample.

Let us introduce the complex number:

$$\Psi(\lambda) = \psi(\lambda) - i\varphi(\lambda). \quad (4.5)$$

Then it follows from Equations 4.2–4.5:

$$\Psi(\lambda) = \frac{i\pi N}{k^2} \Delta S(\lambda), \quad (4.6)$$

where

$$\Delta S(\lambda) = S_2(\lambda) - S_1(\lambda). \quad (4.7)$$

Thus, the problem of the calculation of CD and ORD spectra of a turbid layer is reduced to the calculation of the difference $\Delta S(\lambda)$. The value of $\Delta S(\lambda)$ is small due to the small difference $\Delta m(\lambda) = m_L(\lambda) - m_R(\lambda)$, where m_L and m_R are refractive indices of particles for left-handed and right-handed circularly polarized waves, respectively. It follows due to this fact that:

$$S'(\lambda) = \left.\frac{\partial S}{\partial m}\right|_{m=\bar{m}} = \frac{\Delta S(\lambda)}{\Delta m(\lambda)}. \quad (4.8)$$

Substituting Equation 4.8 into Equation 4.6, one obtains:

$$\Psi(\lambda) = \frac{2\pi}{ik^3 V} S'(\lambda) \Psi_0(\lambda), \qquad (4.9)$$

where V is the average volume of particles,

$$\Psi_0(\lambda) = \psi_0(\lambda) - i\varphi_0(\lambda), \qquad (4.10)$$

$$\psi_0(\lambda) = \frac{\pi C_v}{\lambda} \operatorname{Re}(m_L(\lambda) - m_R(\lambda)), \qquad (4.11)$$

$$\varphi_0(\lambda) = \frac{\pi C_v}{\lambda} \operatorname{Im}(m_L(\lambda) - m_R(\lambda)), \qquad (4.12)$$

and $C_v = NV$ is the volumetric concentration of particles. One can see from Equation 4.9 that the initial spectra $\Psi_0(\lambda) = \psi_0(\lambda) - i\varphi_0(\lambda)$ indeed change and this is due to the spectral variation of the function $\lambda^3 S'(\lambda)$.

Thus, the main task now is to calculate the dependence $S(\lambda)$ and then to consider the derivative $S'(\lambda)$. For this, we need to make some assumptions. In particular, we will assume, as is usually done in the framework of the anomalous diffraction approximation, that the particles are optically soft ($|\bar{m} - 1| \ll 1$) and large ($kd \gg 1$). Then the function $S(\lambda)$ can be found in the framework of the Van de Hulst approximation (Van de Hulst, 1981):

$$S(\lambda) = \frac{k^2}{2\pi} \int_\Sigma (1 - \exp(-i\sigma(p))) \, dp, \qquad (4.13)$$

where Σ is the projection of a particle on the plane perpendicular to the incident light and $\sigma(p) = kl(\bar{m} - 1)\xi(p)$ is the phase shift of a ray with length $l\xi(p)$ where the parameter $l = V/\Sigma$ is introduced for convenience. Equation 4.13 can be applied only to the case of identical particles with a fixed orientation. For more complex cases, one should use values $\langle S(\lambda) \rangle$ averaged over particle dimensions and orientations. It follows from Equation 4.13 for the derivative

$$S'(\lambda) = \frac{ik^3 l}{2\pi} \int_\Sigma \xi(p) \exp[-ik(\bar{m} - 1)l\xi(p)] \, dp. \qquad (4.14)$$

Thus, one can obtain from Equations 4.9 and 4.14:

$$\Psi(\lambda) = H(\lambda) \Psi_0(\lambda), \qquad (4.15)$$

where

$$H(\lambda) = \int_S \xi(q) \exp(-i\rho\xi(q)) \, dq, \qquad (4.16)$$

and $q = p/\Sigma$, $\rho = kl(\bar{m} - 1)$, and S is the unit area. The shape of this unit area coincides with the shape of the particle projection. It follows from Equation 4.15 that:

$$\Psi_0(\lambda) = \frac{\Psi(\lambda)}{H(\lambda)}. \qquad (4.17)$$

Thus, knowledge of the function $H(\lambda)$ allows us to find the intrinsic spectrum $\Psi_0(\lambda)$ of the particles from the measured spectrum $\Psi(\lambda)$. It is known, however, that $H(\lambda) \to 1$ as $\rho \to 0$. This means that

$$\int_S \xi(q) \, dq = 1. \qquad (4.18)$$

This equality holds for arbitrary shapes of particles.

The function $H(\lambda)$ for a given particle size, shape, and orientation can be calculated numerically or in some cases (see below) analytically. For numerical integration, one should find the length $L(q) = l\xi(q)$ of a light ray inside the particle. After that the integration in Equation 4.16 can be performed for a particle of an arbitrary shape. As a matter of fact, a three-dimensional particle is replaced by a two-dimensional amplitude-phase screen in the framework of the Van de Hulst approximation. Thus, the main task is to find the phase shift distribution on that screen.

4.5.3 Cylinders

Let us now consider the case of cylindrical particles. We will assume that cylinders are oriented parallel to the incident light beam. The transverse cross section of a cylinder can be arbitrary. Clearly, for the illumination of the cylinder along the symmetry axis all phase shifts are equal and the equivalent amplitude-phase screen has a constant value of the phase shift. Then it follows: $\xi \equiv 1$ and

$$\rho = kL(\bar{m} - 1), \qquad (4.19)$$

where L is the length of the cylinder. We obtain from Equation 4.16:

$$H(\lambda) = \exp(-i\rho(\lambda)), \qquad (4.20)$$

where we accounted for the fact that $\xi(q) \equiv 1$ in this case. It follows from Equation 4.15

$$\Psi(\lambda) = \Psi_0(\lambda) \exp(-i\rho(\lambda)). \qquad (4.21)$$

Then we have from Equation 4.21 as $\rho(\lambda) \to 0$:

$$\Psi(\lambda) \equiv \Psi_0(\lambda). \qquad (4.22)$$

However, at larger values of $\rho(\lambda)$ considerable differences between spectra $\Psi(\lambda)$ and $\Psi_0(\lambda)$ occur.

In particular, one obtains (see Equation 4.21):

$$\psi(\lambda) = \psi_0(\lambda)\cos\rho(\lambda) - \varphi_0(\lambda)\sin\rho(\lambda), \tag{4.23}$$

$$\varphi(\lambda) = \psi_0(\lambda)\sin\rho(\lambda) + \varphi_0(\lambda)\cos\rho(\lambda) \tag{4.24}$$

if the particles are not absorbing. It follows that at $\rho(\lambda) = 2\pi n (n = 0, 1, 2, 3, \ldots)$:

$$\psi(\lambda) \equiv \psi_0(\lambda), \qquad \varphi(\lambda) \equiv \varphi_0(\lambda). \tag{4.25}$$

However, one obtains at $\rho(\lambda) = \pi(n + \frac{1}{2})$ $(n = 0, 2, 4, 6, 8)$:

$$\psi(\lambda) = -\varphi_0(\lambda), \qquad \varphi(\lambda) = \psi_0(\lambda). \tag{4.26}$$

Equations 4.23 and 4.24 allows the distortion of spectra for arbitrary phase shifts $\rho(\lambda)$ to be considered.

4.5.4 Ellipsoids

Let us consider now the case of ellipsoidal particles in a fixed orientation. The function $\xi(q)$ for this case was found by Lopatin and Sid'ko (1988). It has the following simple form:

$$\xi(q) = \tfrac{3}{2}\sqrt{1-q}. \tag{4.27}$$

It follows from Equations 4.16 and 4.27:

$$H(\lambda) = \frac{3}{2}\int_0^1 \sqrt{1-q}\exp(-i\rho\sqrt{1-q})\,dq, \tag{4.28}$$

where

$$\rho = kd(\bar{m}-1), \qquad d = \tfrac{3}{2}l. \tag{4.29}$$

For spheres the value of d coincides with the diameter. Then, from Equation 4.28 after integration:

$$H(\lambda) = \frac{6i(1-e^{-i\rho})}{\rho^3} + \frac{6e^{-i\rho}}{\rho^2} + \frac{3ie^{-i\rho}}{\rho}. \tag{4.30}$$

We obtain from Equation 4.30 as $\rho \to 0$:

$$H(\lambda) = \left(1 - \frac{3i}{4}\rho - \frac{3}{10}\rho^2\right).$$

One can see again that $H \to 1$ as $\rho \to 0$ as it should be.

The value of $H(\lambda)$ in Equation 4.30 can be presented in the following form

$$H(\lambda) = u(\lambda) - iv(\lambda), \tag{4.31}$$

where

$$u(\lambda) = \frac{3\sin\rho}{\rho} + \frac{6\cos\rho}{\rho^2} - \frac{6\sin\rho}{\rho^3}, \qquad (4.32)$$

$$v(\lambda) = -\frac{3\cos\rho}{\rho} + \frac{6\sin\rho}{\rho^2} - \frac{6(1-\cos\rho)}{\rho^3} \qquad (4.33)$$

and we assume that the particles are non-absorbing. Then we have:

$$\psi(\lambda) = u(\lambda)\psi_0(\lambda) - v(\lambda)\varphi_0(\lambda), \qquad (4.34)$$

$$\varphi(\lambda) = v(\lambda)\psi_0(\lambda) + u(\lambda)\varphi_0(\lambda), \qquad (4.35)$$

which is similar to Equations 4.23 and 4.24. Generally speaking, the linear relationships Equation 4.34 and 4.35 hold for any shape of particles. However, the specific dependencies $u(\lambda)$ and $v(\lambda)$ differ depending on the shape of the particles. We present the functions $u(\lambda)$ and $v(\lambda)$ for cylinders and ellipsoids in Figures 4.19 and 4.20. One can see that the multipliers $H(\lambda)$ for cylinders and ellipsoids differ considerably. In particular, it follows that $u \to 0$, $v \to 0$ as $\rho \to \infty$ for ellipsoids. This is not the case for oriented cylinders. In practice, of course, there is some absorption of light inside particles. This will lead to the damping of the oscillations both for cylinders and spheroids.

For purposes of illustration we present the spectra $\psi(\lambda)$ and $\varphi(\lambda)$ obtained with Equations 4.32–4.35 for different values of the parameter $b = 2\pi d(\bar{m} - 1)$ in Figures 4.21 and 4.22. Note that we neglected the imaginary part of $\bar{m}$ in calculations of the parameter $b \equiv \lambda\rho$. Also we neglected possible variations of b with λ. The spectra $\Delta n(\lambda)$ and $\Delta\chi(\lambda)$, which are close to those of poly-L-glutamic acid, were taken from the PhD thesis of Bohren (1977). Then we used Equations 4.11 and 4.12 to find $\psi_0(\lambda)$ and $\varphi_0(\lambda)$. The concentration of the particles C_v was assumed to be equal to 0.001. Finally, we expressed $\psi_0(\lambda)$ and $\varphi_0(\lambda)$ in degrees. The thickness of the dispersed layer was assumed to be equal to 1 mm. We see that the parameter b, which is proportional to the phase shift of a ray on the maximum dimension of a particle, has paramount importance for the spectra $\psi(\lambda)$ and $\varphi(\lambda)$. At small values of b, the spectra of dispersed layers are very close to those of a molecular solution of poly-L-glutamic acid. This is in a full accordance with data given by Gordon (1972). However, they differ considerably from the initial spectra $\psi_0(\lambda)$ and $\varphi_0(\lambda)$ for large values of b. Generally, we have for large values of $b \to \infty$: $\psi(\lambda) \to 0$ and $\varphi(\lambda) \to 0$. This means that the information contained in the initial spectra is lost due to the light scattering process. However, at small and intermediate values of b the situation is not so hopeless (see Figures 4.21, 4.22) and information on the arrangement of chemical groups for substances condensed inside small particles, which is responsible for the particular shapes of the spectra $\psi_0(\lambda)$ and $\varphi_0(\lambda)$, can be retrieved in principle. For this, one needs to have additional information on the shape, size, and internal structure of the particles under study.

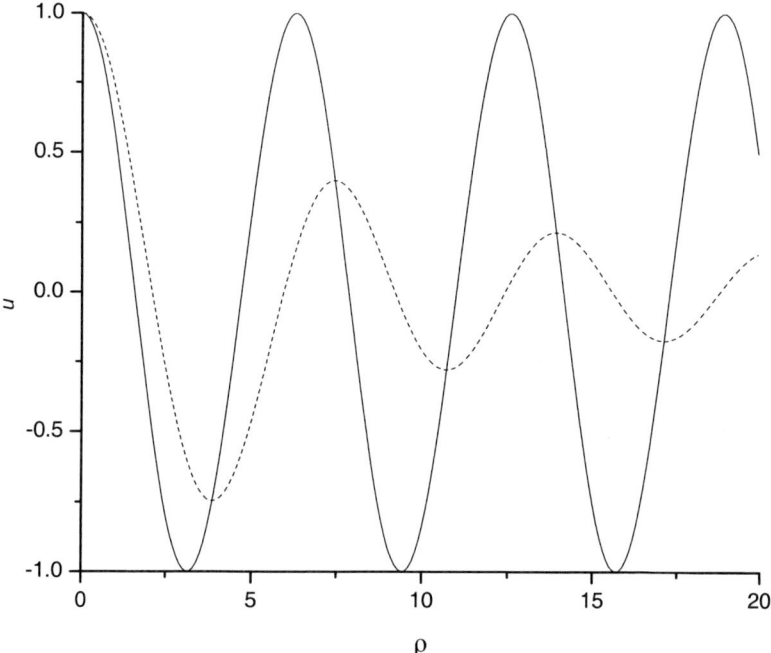

Figure 4.19. The dependence of u on ρ (solid line – cylinders, dashed line – ellipsoids).

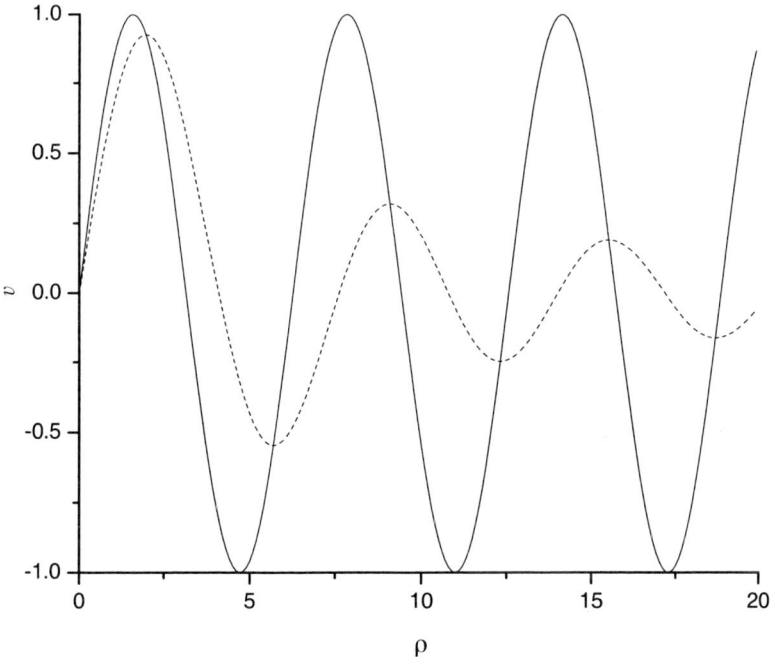

Figure 4.20. The dependence of v on ρ (solid line – cylinders, dashed line – ellipsoids).

192 Environmental polarimetry [Ch. 4]

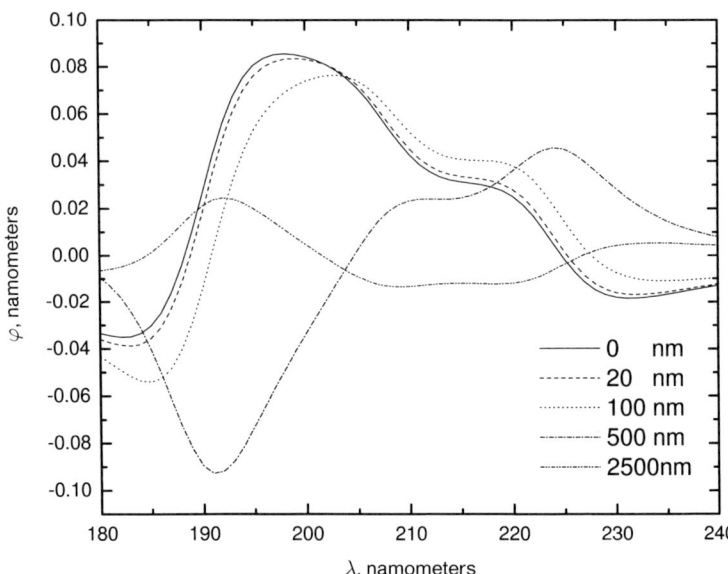

Figure 4.21. The ORD spectrum for various values of $b = 2\pi d(\bar{m} - 1)$ equal to 0, 20, 100, 500 and 2500 μm. The thickness of a dispersed layer is equal to 1 mm and the volumetric concentration of particles is equal to 0.001.

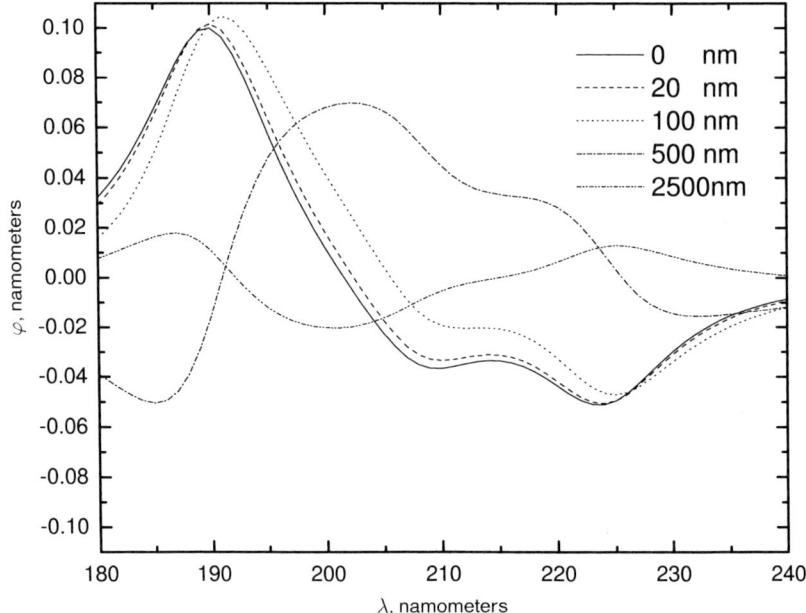

Figure 4.22. The same as in Figure 4.21 but for the CD spectrum.

4.6 FURTHER READING

Bohren, C. F. (1974) Light scattering by an optically active sphere. *Chem. Phys. Lett.*, **29**, 458–462.
Bohren, C. F. (1975a) Light scattering by optically active particles. PhD thesis, University of Arizona, Tucson, AZ.
Bohren, C. F. (1975b) Scattering of electromagnetic waves by an optically active spherical shell. *J. Chem. Phys.*, **62**, 1566–1571.
Bohren, C. F. (1977) Circular dichroism and optical rotatory dispersion spectra of arbitrarily shaped optically active particles. *J. Theor. Biol.*, **65**, 755–767.
Bohren, C. F. (1978) Scattering of electromagnetic waves by an optically active cylinder, *J. Colloid Interface Sci.*, **66**, 105–109.
Bohren, C. F. and Huffman, D. R. (1983) *Absorption and Scattering of Light by Small Particles*. Wiley, New York.
Breon, F.-M. and Goloub, P. (1998) Cloud droplet effective radius from space-borne polarization measurements. *Geophys. Res. Let.*, **25**, 1879–1883.
Chandrasekhar, S. (1950) *Radiative Transfer*. Oxford University Press, Oxford.
Chepfer, H., Brogniez, G. and Fouquart, Y. (1998) Cirrus clouds microphysical properties deduced from POLDER observations. *JQSRT*, **60**, 375–390.
Chepfer, H., Brogniez, G., Goloub, P., Breon, F. M. and Flamant, P. H. (1999) Observations of horizontally oriented ice crystals in cirrus clouds with POLDER-1/ADEOS-1. *JQSRT*, **63**, 521–543.
Chepfer, H., Goloub, P., Riedi, J., de Haan, J. F., Hovenier, J. W. and Flamant, P. H. (2001) Ice crystal shapes in cirrus clouds derived from POLDER/ADEOS-1. *J. Geophys. Res. D*, **106**, 7955–7966.
Coffeen, D. L. (1979) Polarization and scattering characteristics in the atmospheres of Earth, Venus, and Jupiter. *J. Opt. Soc. America*, **69**, 1051–1064.
Coulson, K. L., (1988) *Polarization and Intensity of Light in Atmosphere*. Deepak, Hampton, VA.
Deirmendjian, A. (1969) *Electromagnetic Scattering on Spherical Polydispersions*. Elsevier, Amsterdam.
Deschamps, P.-Y., Breon, F. M., Leroy, M., Podaire, A., Bricaud, A., Buriez, J. C. et al. (1994) The POLDER Mission: Instrument characteristics and scientific objectives. *IEEE Trans.*, **GE 32**, 598–614.
Dolginov A. Z., Gnedin, Yu. N. and Silant'ev, N. A. (1995) *Propagation and Polarization of Radiation in Cosmic Media*. Gordon & Breach, Amsterdam.
de Haan, J. F. (1987) *Effects of Aerosols on the Brightness and Polarization of Cloudless Planetary Atmospheres*. PhD thesis, Free University of Amsterdam, Enschede, The Netherlands.
Gehrels, T. (ed.) (1974) *Planets, Stars and Nebulae Studied with Photopolarimetry*. The University of Arizona Press, Tuscon, AZ.
Gorchakov, G. I. (1966) Light scattering matrices of atmospheric air in a boundary layer. *Izvestiya, Atmospheric and Oceanic Physics*, **2**, 595–605.
Gorchakov, G. I. (1971) On the degree of the polarization coherence of light scattered by atmospheric air. *Izvestiya, Atmospheric and Oceanic Physics*, **7**, 224–227.
Gordon, D. J. (1972) Mie scattering by optically active particles. *Biochemistry*, **11**, 413–420.
Hansen, J. E. (1971) Multiple scattering of light in planetary atmospheres clouds. Part II: Sunlight reflected by terrestrial water clouds. *J. Atmos. Sci.*, **28**, 1400–1426.

Hapke, B. (1993) *Theory of the Reflectance and Emmitance Spectroscopy*. Cambridge University Press, Cambridge, UK.

Ivanoff, A. and Waterman, T. H. (1958a) Elliptical polarization of submarine illumination. *J. Mar. Res.*, **16**, 255–282.

Ivanoff, A. and Waterman, T. H. (1958b) Factors, mainly depth and wavelength, affecting the degree of underwater light polarization. *J. Mar. Res.*, **16**, 283–307.

Kadyshevich, E. A., Lubovtseva, E. S. and Rozenberg, G. V. (1974) Matrices of light scattering by waters of the Pacific and Atlantic oceans. *Izv. AN SSSR, Fizika Atmos. Okeana*, **12**, 186–195.

Kokhanovsky, A. A. (2001) *Light Scattering Media Optics: Problems and Solutions*. Springer–Praxis, Chichester, UK.

Kokhanovsky, A. A. and Weichert, R. (2002) Determination of the droplet effective size and optical depth of cloudy media from polarimetric measurements: Theory. *Appl. Optics*, **41**, 3650–3658.

Leroux, C., Lenoble, J., Brogniez, G., Hovenier, J. W. and de Haan, J. F. (1997) A model for the bidirectional polarized reflectance of snow. *J. Quant. Spectr. Radiat. Transfer*, **61**, 273–285.

Leroy, J.-L. (2000) *Polarization of light and astronomical observation*. Gordon and Breach Science Publishers, Amsterdam.

Liou, K. N. (1992) *Radiation and Cloud Processes in the Atmosphere*. Oxford University Press, Oxford, UK.

Liou, K. N., Takano, Y., Yang, P. and Gu, Y. (2002) Radiative transfer in Cirrus clouds: Light scattering and spectral information. *Cirrus* (edited by K. Lynch, K. Sassen, D. O'C. Starr and G. Stephens, pp. 265–296). Oxford University Press, Oxford, UK.

Lopatin, V. N. and Sid'ko, F. Ya. (1988) *Introduction to Optics of Cell Suspensions*. Nauka, Moscow.

Masuda, K., Sasaki, M., Takashima, T. and Ishida, H. (1999) Use of polarimetric measurements of the sky over the Ocean for spectral optical thickness retrievals. *J. Atmos. and Oceanic Techn.*, **16**, 846–859.

Masuda, K., Ishimoto, H. and Takashima T. (2002a) Dependence of the spectral aerosol optical thickness retrieval from space on measurement errors and model parameters. *Int. J. Remote Sensing*, **23**, 3835–3851.

Masuda, K., Ishimoto, H. and Takashima, T. (2002b) Retrieval of cirrus optical thickness and ice-shape information using total and polarized reflectance from satellite measurements. *J. Quant. Spectr. Radiat. Transfer*, **75**, 39–51.

McCartney, E. J. (1977) *Optics of the Atmosphere*. Wiley, New York.

Mishchenko, M. I., et al. (1996) Sensitivity of cirrus cloud albedo, bi-directional reflectance function and optical thickness retrieval accuracy to ice particle shape. *J. Geophys. Res.*, **102**, 16831–16847.

Mobley, C. D. (1994) *Light and Water: Radiative Transfer in Natural Waters*. Academic Press, San Diego.

Sekera, Z. (1956) Recent developments in the study of the polarization of sky light. *Adv. in Geophys.*, **3**, 43–104.

Shifrin, K. S. (1988) *Introduction into Ocean Optics*. Gidrometeoizdat, Leningrad.

Stam, D. M., Aben, I. and Helderman, F. (2002): Skylight polarization spectra: Numerical simulation of the Ring effect. *J. Geophys. Res. D*, **107**, 10.1.1029/2001JD000951.

Stramski, D., Bricaud, A. and Morel, A. (2001) Modeling of the inherent optical properties of the ocean based on the detailed composition of the planktonic community. *Appl. Optics* **40**, 2929–2945.

Voss, K. J. and Fry, E. S. (1984) Measurement of the Mueller matrix for ocean water. *Appl. Optics*, **23**, 4427–4436.

Van de Hulst, H. C. (1981) *Light Scattering by Small Particles*. Dover, New York.

Von Hoyningen-Huene, W., Freitag, M. and Burrows, J. P. (2003) Retrieval of aerosol optical thickness over land surfaces from top-of-atmosphere radiance. *J. Geophys. Res. D*, **108**, 10.1029/2001JDOO2018.

WCP-112 (1986) *A Preliminary Cloudless Standard Atmosphere for Radiation Computation*, pp. 53. World Meteorological Organization, Geneva.

Zhang, T. and Gordon, H. R. (1997) Retrieval of elements of the columnar aerosol scattering phase matrix from polarized sky radiance over the ocean: Simulations. *Appl. Opt.*, **36**, 7948–7959.

Zhao, F., Gong, Z., Hu, H., Tanaka, M. and Hayasaka, T. (1997) Simultaneous determination of the aerosol complex index of refraction and size distribution from scattering measurements of polarized light. *Appl. Opt.*, **36**, 7992–8001.

Appendix 1

The tensor radiative transfer equation

The propagation of light in turbid media is often described in the framework of the vector radiative transfer equation. This equation was formulated by Chandrasekhar (1951) and in the more general case of anisotropic media by Rozenberg (1955). It describes the change of the Stokes vector components during light propagation in scattering media. The components of the Stokes vector $\vec{S}(I, Q, U, V)$ can be related to the electric vector by the following equations:

$$I = |E_1|^2 + |E_2|^2, \quad Q = |E_1|^2 - |E_2|^2, \quad U = 2\text{Re}\,(E_1 E_2^*), \quad V = -2\text{Im}\,(E_1 E_2^*),$$

(A1.1)

where values of E_1 and E_2 are components of the electric field in the plane perpendicular to the direction of propagation. These equations coincide with Equations 1.9–1.12.

One can see that the Stokes vector is defined in the framework of the special coordinate system, which is attached to the direction of light propagation. This direction is changed many times during the process of light scattering in a disperse medium. This feature is accounted for with the use of special rotation matrices. It is of general importance to have the radiative transfer equation in covariant tensor form. The covariant methods (not dependent on the specific coordinate system) were introduced into the optics of anisotropic and gyrotropic media by Fedorov (1958, 1976). They have allowed for the rapid progress in the field of uniform anisotropic media.

Let us introduce the light beam tensor (Fedorov, 1976):

$$F = \sum_s \vec{E}^{(s)} \vec{E}^{(s)*}, \quad \vec{n}\vec{E}^{(s)} = \vec{n}\vec{E}^{(s)*} = 0, \quad (A1.2)$$

where $\vec{E} = E_1\vec{x} + E_2\vec{y} + E_3\vec{z}$ is the electric vector, $\sum_s$ means the summation on all incoherent simple waves in a beam, and the vector $\vec{n}$ defines the direction of propagation. The dyadic notation $F = \vec{E}\vec{E}^*$ means:

$$F = \begin{pmatrix} E_1 E_1^* & E_1 E_2^* & E_1 E_3^* \\ E_2 E_1^* & E_2 E_2^* & E_2 E_3^* \\ E_3 E_1^* & E_3 E_2^* & E_3 E_3^* \end{pmatrix}. \tag{A1.3}$$

In the framework of the coordinate system attached to the direction of propagation $\vec{n}\|\vec{z}$, one obtains $E_3 = 0$ and the tensor F reduces to the density matrix

$$\rho = \begin{pmatrix} |E_1|^2 & E_1 E_2^* \\ E_2 E_1^* & |E_2|^2 \end{pmatrix}. \tag{A1.4}$$

The elements of the matrix ρ can be expressed in terms of the components of the Stokes vector $\vec{S}$. Thus, both representations (A1.1) and (A1.4) are coordinate dependent. The radiative transfer equation for the density matrix ρ is given by Dolginov et al. (1995).

Fedorov (1976) introduced the following invariants of the light beam tensor F:

$$I = F_t, \qquad K = (F^2)_t, \qquad M = i(n^\times F)_t, \qquad L = (FF^*)_t, \tag{A1.5}$$

where t means the trace (e.g., $F_t = |E_1|^2 + |E_2|^2 + |E_3|^2$) and $n^\times$ is the tensor with components

$$n^\times_{abc} = e_{abc} n_b \tag{A1.6}$$

or

$$n^\times = \begin{pmatrix} 0 & -n_3 & n_2 \\ n_3 & 0 & -n_1 \\ -n_2 & n_1 & 0 \end{pmatrix} \tag{A1.7}$$

where e_{abc} is the well known Levi–Civita tensor.

The first invariant describes the intensity of the light beam I. The invariant M coincides with the V component of the Stokes vector. Thus, the first and last components of the Stokes vectors (see Equation A1.1) are coordinate independent. This is not the case for components Q and U. The ratio K/I^2 determines the degree of polarization

$$p = \sqrt{\frac{2K}{I^2} - 1} \tag{A1.8}$$

and the invariant L can be used to find the semi-axes of the polarization ellipse:

$$a = \sqrt{\frac{I + \sqrt{L}}{2}}, \qquad b = \sqrt{\frac{I - \sqrt{L}}{2}}. \tag{A1.9}$$

One obtains at $L = 0$:

$$a = b = \sqrt{\frac{I}{2}}. \tag{A1.10}$$

This is the case of circularly polarized light.

The components (A1.1) can be found from invariants of the tensor F. Thus, the light beam tensor can be used for the complete description of the polarization characteristics of a light beam.

Let us find the transfer equation for the tensor F in a random medium. The transformation of the electric field $\vec{E}$ is due to the linear interaction of the light beam with a scatterer. It can be described by the interaction tensor γ:

$$\vec{E}' = \gamma \vec{E}. \tag{A1.11}$$

It follows from Equations A1.2 and A1.11:

$$F' = \sum_s \vec{E}'^{(s)} \vec{E}'^{(s)*} = \sum_s \gamma \, \vec{E}^{(s)} \vec{E}^{(s)*} \gamma^+ = \gamma F \gamma^+, \tag{A1.12}$$

where $\gamma^+ = \tilde{\gamma}^*$ and $\tilde{\gamma}$ is the transpose tensor.

First of all it should be pointed out that due to the scattering process the incident plane wave transforms to the spherical wave and the interaction matrix can be written as

$$\gamma = f(\vec{n}', \vec{n}) \frac{e^{ikr}}{r}, \tag{A1.13}$$

where $\vec{n}'$ and $\vec{n}$ are directions of the incident and scattering waves respectively, r is the distance to the observation point and k is the wave number. It follows from Equations A1.12 and A1.13:

$$F(\vec{n}) = r^{-2} f(\vec{n}', \vec{n}) F(\vec{n}') f^+(\vec{n}', \vec{n}). \tag{A1.14}$$

The contribution of incoherent scattering to the path dn from all directions to the direction $\vec{n}$ can be presented as follows:

$$dF^{(1)}(\vec{n}) = N \, dn \int d\Omega_{\vec{n}'} f(\vec{n}', \vec{n}) F(\vec{n}') f^+(\vec{n}', \vec{n}), \tag{A1.15}$$

where N is the number of scatterers in a unit volume, $d\Omega_{\vec{n}'}$ is the solid angle along the direction $\vec{n}'$. The removal of photons from the direction $\vec{n}$ on the path dn due to the extinction process can be described by the operator $\hat{L}$:

$$dF^{(2)}(\vec{n}) = -N\hat{L}F(\vec{n}) \, dn. \tag{A1.16}$$

According to the energy conservation law it follows from Equations A1.15 and A1.16:

$$\frac{dF(\vec{n})}{dn} = -N\hat{L}F(\vec{n}) + N \int d\Omega_{\vec{n}'} f(\vec{n}', \vec{n}) F(\vec{n}') f^+(\vec{n}', \vec{n}) + B(\vec{n}), \tag{A1.17}$$

where the term $B(\vec{n})$ describes internal emitting sources. This is the radiative transfer equation written in the framework of the coordinate-free approach.

Appendix 1

Let us now determine the operator $\hat{L}$. The tensor F can be presented as follows:

$$F(\vec{n}) = F_c(\vec{n})\delta(\vec{n} - \vec{n}_o) + F_d(\vec{n}), \qquad (A1.18)$$

where $\delta(\vec{n} - \vec{n}_o)$ is the delta function, $F_c(\vec{n})$ and $F_d(\vec{n})$ are the coherent and diffused (incoherent) parts of the tensor F. The direction of the vector $\vec{n}_o$ coincides with the direction of incidence. We have from Equations A1.17 and A1.18:

$$\dot{F}_c(\vec{n}_0) + \dot{F}_d(\vec{n}) = -N\hat{L}F_c(\vec{n}_0) - N\hat{L}F_d(\vec{n}) + N\int d\Omega_{\vec{n}'} f(\vec{n}', \vec{n}) F_d(\vec{n}') f^+(\vec{n}', \vec{n})$$

$$+ Nf(\vec{n}_0, \vec{n})F_c(\vec{n}_0) f^+(\vec{n}_0, \vec{n}) + B(\vec{n}), \qquad (A1.19)$$

where $\dot{F}(\vec{n}) \equiv \dfrac{dF}{dn} = (\vec{n} \cdot \vec{\nabla})F$.

Thus, one can obtain the following equations:

$$\dot{F}_c(\vec{n}_0) = -N\hat{L}F_c(\vec{n}_0) \qquad (A1.20)$$

for the coherent component and

$$\dot{F}_d(\vec{n}) = -N\hat{L}F_d(\vec{n}) + N\int d\Omega_{\vec{n}'} f(\vec{n}', \vec{n}) F_d(\vec{n}') f^+(\vec{n}', \vec{n}) + B_0(\vec{n}) + B(\vec{n}), \qquad (A1.21)$$

for the diffused light. The tensor

$$B_0(\vec{n}) = Nf(\vec{n}_0, \vec{n}) F_c(\vec{n}_0) f^+(\vec{n}_0, \vec{n}) \qquad (A1.22)$$

describes single scattering of the incident light and can be obtained from the solution of Equation A1.20.

It is well known that the coherent field $\vec{E}_c$ propagating in the direction $\vec{n}_0$ satisfies the equation (Ishimaru and Yeh, 1984):

$$\dot{\vec{E}}_c(\vec{n}_0) = i\lambda Nf(\vec{n}_0, \vec{n}_0)\vec{E}_c(\vec{n}_0), \qquad (A1.23)$$

where λ is the wavelength.

From Equation A1.23 and the definition of the coherent component of the light beam tensor

$$F_c = \vec{E}_c \vec{E}_c^* \qquad (A1.24)$$

it follows that:

$$\dot{F}_c = \dot{\vec{E}}_c \vec{E}_c^* + \vec{E}_c \dot{\vec{E}}_c^* = i\lambda N(f\vec{E}_c\vec{E}_c^* - \vec{E}_c f^* \vec{E}_c^*) \qquad (A1.25)$$

or

$$\dot{F}_c = i\lambda N(fF_c - F_c f^+), \qquad (A1.26)$$

where we have omitted the arguments. One can see, comparing Equation A1.20 and A1.26, that:

$$\hat{L}|\psi\rangle = -i\lambda(f|\psi\rangle - \langle\psi|f^+), \qquad (A1.27)$$

where we have used the Dirac's notation.

Finally, we obtain the covariant radiative transfer equation (CRTE) for diffused light (see Equations A1.21 and A1.22):

$$(\vec{n} \cdot \vec{\nabla})F_d(\vec{r}, \vec{n}) = i\lambda N[f(\vec{n}, \vec{n})F_d(\vec{r}, \vec{n}) - F_d(\vec{r}, \vec{n})f^+(\vec{n}, \vec{n})]$$

$$+ N \int d\Omega_{\vec{n}'} f(\vec{n}', \vec{n})F_d(\vec{r}, \vec{n}')f^+(\vec{n}', \vec{n})$$

$$+ Nf(\vec{n}_0, \vec{n})F_c(\vec{r}, \vec{n}_0)f^+(\vec{n}_0, \vec{n}) + B(\vec{n}), \qquad (A1.28)$$

where $\vec{r}$ is the radius vector of the observation point. The value of $F_c(\vec{n}_0)$ in Equation A1.28 is determined from Equation A1.26:

$$(\vec{n} \cdot \vec{\nabla})F_c = i\lambda N(fF_c - F_c f^+) \qquad (A1.29)$$

or

$$F_c(\vec{n}_0) = \exp\{-N\hat{L}\}F_0(\vec{n}_0), \qquad (A1.30)$$

where $F_0(\vec{n}_0)$ is the beam tensor for the incident light. Equation A1.30 is the generalized Bouguer law.

The boundary conditions for Equation A1.28 state that there is no diffused light coming into a scattering convex medium from outside:

$$F_d(\vec{r}_0, \vec{n}) = 0 \qquad (\text{at } \vec{n}\vec{l} < 0), \qquad (A1.31)$$

where $\vec{l}$ is the unit vector normal to the boundary in the outward direction at the point with the radius vector $\vec{r}_0$.

Note that the general CRTE can be obtained from Equations A1.17 and A1.27:

$$(\vec{n} \cdot \vec{\nabla})F(\vec{r}, \vec{n}) = i\lambda N[f(\vec{n}, \vec{n})F(\vec{r}, \vec{n}) - F(\vec{r}, \vec{n})f^+(\vec{n}, \vec{n})]$$

$$+ N \int d\Omega_{\vec{n}'} f(\vec{n}', \vec{n})F(\vec{r}, \vec{n})f^+(\vec{n}', \vec{n}) + B(\vec{n}), \qquad (A1.32)$$

with the boundary condition

$$F(\vec{r}_0, \vec{n}) = F_0(\vec{r}_0, \vec{n}) \qquad (\text{at } \vec{n}\vec{l} < 0), \qquad (A1.33)$$

where $\vec{r}_0$ is the radius vector on the point at the boundary of the scattering medium.

Equation A1.32 reduces to the well known radiative transfer equation for the density matrix (Gnedin et al., 1995) at $\vec{n} \| \vec{z}$ (see Equation A1.4). It follows from Equation A1.32 in the framework of the scalar approximation:

$$(\vec{n} \cdot \vec{\nabla})I(\vec{r}, \vec{n}) = -\sigma_{ext} I(\vec{r}, \vec{n}) + N \int d\Omega_{\vec{n}'} \sigma_{sca}(\vec{n}', \vec{n}) I(\vec{r}, \vec{n}') + B(\vec{n}), \qquad (A1.34)$$

where

$$\sigma_{ext} = 2\lambda N \operatorname{Im} (f(\vec{n}, \vec{n})), \qquad (A1.35)$$

$$\sigma_{sca}(\vec{n}', \vec{n}) = |f(\vec{n}', \vec{n})|^2. \qquad (A1.36)$$

This is the well-known scalar radiative transfer equation.

Appendix 1

To solve Equation A1.32 one needs to know the 3×3 interaction tensor $f(\vec{n}, \vec{n}')$ of the scattering medium. It depends on the size of the scatterers, their shape, and the dielectric tensors of particles' substances, etc. This tensor can vary with location inside a medium. Many particles can contribute to the value of $f(\vec{n}', \vec{n})$. Thus, the value of $f(\vec{n}, \vec{n}')$ in previous equations is the average value of the interaction tensor for the ensemble of scatterers. It can be expressed by simple integrals in the framework of the Born approximation (Ishimaru, 1978), namely

$$f(\vec{n}, \vec{n}') = \frac{\pi V}{\lambda^2}[A - \vec{n}'(\vec{n}' \cdot A)], \qquad (A1.37)$$

where V is the volume of a scatterer and

$$A = \frac{1}{V}\int_V \left(\frac{\varepsilon(\vec{r})}{\varepsilon_m} - I\right)e^{-i\vec{q}\vec{r}}\, d^3\vec{r}. \qquad (A1.38)$$

Here $\varepsilon(\vec{r})$ is the dielectric tensor of the scatterer, ε_m is the dielectric permittivity of the host medium, I is the unit matrix and $\vec{q} = k(\vec{n}' - \vec{n})$ with $k = 2\pi/\lambda$. To derive Equation (A1.37) it was assumed that the electric field inside a weak scatterer is equal to the electric field of the incident wave.

FURTHER READING

Chandrasekhar, S. (1951) *Radiative Transfer*. Oxford University Press, Oxford, UK.
Dolginov, A. Z., Gnedin, Yu. N. and Silant'ev, N. A. (1995) *Propagation and Polarization of Radiation in Cosmic Media*. Gordon & Breach, Amsterdam.
Fedorov, F. I. (1958) *Optics of Anisotropic Media*. Academy of Sciences of Belarus, Minsk.
Fedorov, F. I. (1976) *Theory of Gyrotropy*. Science and Technology, Minsk.
Gnedin, Yu. N., Silant'ev, N. A. and Dolginov, A. Z. (1995) *Propagation and Polarization of Radiation in Cosmic Media*, Gordon and Breach, New York.
Ishimaru, A. (1978) *Wave Propagation and Scattering in Random Media*. Academic Press, New York.
Ishimaru, A. and Yeh, C. W. (1984) Matrix representation of the vector radiative transfer theory for randomly distributed nonspherical particles. *J. Opt. Soc. America*, **A1**, 359–364.
Rozenberg, G. V. (1955) Stokes vector-parameter. *Uspekhi Fiz. Nauk*, **56**, 77–110.
Sobolev, V. V. (1956) *Radiative Transfer in Stellar and Planetary Atmospheres*. Gostekhteorizdat, Moscow.

Appendix 2

Jones matrices, Mueller matrices, and Stokes vectors

Table A2.1. Jones and Mueller matrices.

Polarizer, phase-shifter (plate) or rotator	Jones matrix	Mueller matrix
Linear polarizer parallel to OX	$\begin{pmatrix} 1 & 0 \\ 0 & 0 \end{pmatrix}$	$\dfrac{1}{2}\begin{pmatrix} 1 & 1 & 0 & 0 \\ 1 & 1 & 0 & 0 \\ 0 & 0 & 0 & 0 \\ 0 & 0 & 0 & 0 \end{pmatrix}$
Linear polarizer parallel to OY	$\begin{pmatrix} 0 & 0 \\ 0 & 1 \end{pmatrix}$	$\dfrac{1}{2}\begin{pmatrix} 1 & -1 & 0 & 0 \\ -1 & 1 & 0 & 0 \\ 0 & 0 & 0 & 0 \\ 0 & 0 & 0 & 0 \end{pmatrix}$
Linear polarizer oriented at $+45°$	$\dfrac{1}{2}\begin{pmatrix} 1 & 1 \\ 1 & 1 \end{pmatrix}$	$\dfrac{1}{2}\begin{pmatrix} 1 & 0 & 1 & 0 \\ 0 & 0 & 0 & 0 \\ 1 & 0 & 1 & 0 \\ 0 & 0 & 0 & 0 \end{pmatrix}$
Linear polarizer $\theta = (OX, OY)$	$\begin{pmatrix} \cos^2\theta & \sin\theta\cos\theta \\ \sin\theta\cos\theta & \sin^2\theta \end{pmatrix}$	$\dfrac{1}{2}\begin{pmatrix} 1 & \cos 2\theta & \sin 2\theta & 0 \\ \cos 2\theta & \cos^2\theta & \cos 2\theta\sin 2\theta & 0 \\ \sin 2\theta & \cos 2\theta\sin 2\theta & \sin^2 2\theta & 0 \\ 0 & 0 & 0 & 0 \end{pmatrix}$
Left-handed circular polarizer	$\dfrac{1}{2}\begin{pmatrix} 1 & -i \\ +i & 1 \end{pmatrix}$	$\dfrac{1}{2}\begin{pmatrix} 1 & 0 & 0 & 1 \\ 0 & 0 & 0 & 0 \\ 0 & 0 & 0 & 0 \\ 1 & 0 & 0 & 1 \end{pmatrix}$

Appendix 2

	Jones matrix	Mueller matrix
Right-handed circular polarizer	$\dfrac{1}{2}\begin{pmatrix} 1 & +i \\ -i & 1 \end{pmatrix}$	$\dfrac{1}{2}\begin{pmatrix} 1 & 0 & 0 & -1 \\ 0 & 0 & 0 & 0 \\ 0 & 0 & 0 & 0 \\ -1 & 0 & 0 & 1 \end{pmatrix}$
Depolarizer	—	$\begin{pmatrix} 1 & 0 & 0 & 0 \\ 0 & 0 & 0 & 0 \\ 0 & 0 & 0 & 0 \\ 0 & 0 & 0 & 0 \end{pmatrix}$
Half-wave plate	$\begin{pmatrix} 1 & 0 \\ 0 & -1 \end{pmatrix}$	$\begin{pmatrix} 1 & 0 & 0 & 0 \\ 0 & 1 & 0 & 0 \\ 0 & 0 & -1 & 0 \\ 0 & 0 & 0 & -1 \end{pmatrix}$
Quarter-wave plate	$\begin{pmatrix} 1 & 0 \\ 0 & -1 \end{pmatrix}$	$\begin{pmatrix} 1 & 0 & 0 & 0 \\ 0 & 1 & 0 & 0 \\ 0 & 0 & 0 & 1 \\ 0 & 0 & -1 & 0 \end{pmatrix}$
Phase plate (ϕ arbitrary)	$\begin{pmatrix} e^{-i\phi/2} & 0 \\ 0 & e^{i\phi/2} \end{pmatrix}$	$\begin{pmatrix} 1 & 0 & 0 & 0 \\ 0 & 1 & 0 & 0 \\ 0 & 0 & \cos\phi & \sin\phi \\ 0 & 0 & -\sin\phi & \cos\phi \end{pmatrix}$
Rotator with angle α	$\begin{pmatrix} \cos\alpha & -\sin\alpha \\ \sin\alpha & \cos\alpha \end{pmatrix}$	$\begin{pmatrix} 1 & 0 & 0 & 0 \\ 0 & \cos 2\alpha & -\sin 2\alpha & 0 \\ 0 & \sin 2\alpha & \cos 2\alpha & 0 \\ 0 & 0 & 0 & 1 \end{pmatrix}$

Table A2.2. Stokes vectors.

Polarization state	Stokes vector
Horizontally polarized light	$\begin{pmatrix} 1 \\ 1 \\ 0 \\ 0 \end{pmatrix}$
Vertically polarized light	$\begin{pmatrix} 1 \\ -1 \\ 0 \\ 0 \end{pmatrix}$
Linearly polarized light (+45°)	$\begin{pmatrix} 1 \\ 0 \\ 1 \\ 0 \end{pmatrix}$
Linearly polarized light (−45°)	$\begin{pmatrix} 1 \\ 0 \\ -1 \\ 0 \end{pmatrix}$
Right-handed circularly polarized light	$\begin{pmatrix} 1 \\ 0 \\ 0 \\ 1 \end{pmatrix}$
Left-handed circularly polarized light	$\begin{pmatrix} 1 \\ 0 \\ 0 \\ -1 \end{pmatrix}$
General case	$\begin{pmatrix} 1 \\ \cos(2\psi)\cos(2\varphi) \\ \sin(2\psi)\cos(2\varphi) \\ \sin(2\varphi) \end{pmatrix}$

FURTHER READING

Collett, E. (1993) *Polarized Light: Fundamentals and Applications.* Marcel Dekker, New York.
Huard, S. (1997) *Polarization of Light.* Wiley, New York.

Appendix 3

The system of linear differential equations

The task of this Appendix is to solve the system of differential equations:

$$\cos\vartheta \frac{d\vec{I}}{dZ} = -\hat{\sigma}_{ext}\vec{I} + \vec{W}, \qquad (A3.1)$$

which describe coherent wave propagation through a plane-parallel scattering slab. We will use the following form of this equation:

$$\frac{d\vec{I}}{dx} = -\hat{\sigma}_{ext}\vec{I} + \vec{W}, \qquad (A3.2)$$

where $x = Z/\cos\vartheta$. Vectors $\vec{I}$, $\vec{W}$, and the matrix $\hat{\sigma}_{ext}$ are defined in Chapter 2. In particular, it follows for the components of the vector $\vec{W}$:

$$W_p(x) = \sigma^d_{sca\,ps} M_{sk}(x) J_k,$$

where the matrix $\hat{\sigma}^d_{sca\,pq}$ and the vector $\vec{J}$ are also defined in Chapter 2. They do not depend on x. Also we have (see Chapter 2):

$$M_{sk}(x) = \tfrac{1}{2} Tr(\hat{\sigma}_s \hat{C}(x) \hat{\sigma}_k \hat{C}^+(x)),$$

where

$$\ddot{C}(x) = \hat{D}_1 \exp(-\gamma_1 x) + \hat{D}_2 \exp(-\gamma_2 x).$$

The matrices $\hat{D}_1$ and $\hat{D}_2$ do not depend on x and given in Chapter 2. Therefore:

$$M_{sk}(x) = a_{sk}\exp(-\Lambda_1 vx) + b_{sk}\exp(-\Lambda_2 vx) + c_{sk}\exp(-\Lambda_3 vx) + d_{sk}\exp(-\Lambda_4 vx),$$

where $v = \cos\vartheta/\cos\vartheta_0$ and

$$a_{sk} = \tfrac{1}{2}Tr(\hat{\sigma}_s \hat{D}_1(x)\hat{\sigma}_k D_1^+(x)),$$

$$b_{sk} = \tfrac{1}{2}Tr(\hat{\sigma}_s \hat{D}_1(x)\hat{\sigma}_k D_2^+(x)),$$

$$c_{sk} = \tfrac{1}{2}Tr(\hat{\sigma}_s \hat{D}_1^+(x)\hat{\sigma}_k D_2(x)),$$

$$d_{sk} = \tfrac{1}{2}Tr(\hat{\sigma}_s \hat{D}_2(x)\hat{\sigma}_k D_2^+(x)),$$

$$\Lambda_1 = \gamma_1 + \gamma_1^*, \qquad \Lambda_2 = \gamma_1 + \gamma_2^*, \qquad \Lambda_3 = \gamma_1^* + \gamma_2, \qquad \Lambda_4 = \gamma_2 + \gamma_2^*.$$

Let us introduce the matrix $\hat{H}$ with the following property:

$$\hat{\Lambda} = \hat{H}^{-1}\hat{\sigma}_{ext}\hat{H},$$

where $\hat{\Lambda}$ is the diagonal matrix, having the form:

$$\begin{pmatrix} \Lambda_1 & 0 & 0 & 0 \\ 0 & \Lambda_2 & 0 & 0 \\ 0 & 0 & \Lambda_3 & 0 \\ 0 & 0 & 0 & \Lambda_4 \end{pmatrix}.$$

This matrix can easily be found by calculating the eigenvalues of the extinction matrix $\hat{\sigma}_{ext}$, which yield the columns of the matrix $\hat{H}$. Let us multiply Equation A3.2 by $\hat{H}^{-1}$. Then it follows:

$$\hat{H}^{-1}\frac{d\vec{I}}{dx} = -\hat{H}^{-1}\hat{\sigma}_{ext}\vec{I} + \hat{H}^{-1}\vec{W}$$

or

$$\frac{d\vec{Y}}{dx} = -\hat{\Lambda}\vec{Y} + \vec{\beta},$$

where $\vec{Y} = \hat{H}^{-1}\vec{I}$, and $\vec{\beta} = \hat{H}^{-1}\vec{W}$. The last equation can be written in the following form:

$$\frac{dY_i}{dx} = -\Lambda_i Y + \beta_i.$$

Therefore, we have four independent linear differential equations with solutions:

$$Y_i(x) = \int_\alpha^x \exp(-\Lambda_i(x-s))\beta_i(s)\,ds,$$

where we accounted for the fact that there is no diffuse light incident on the boundaries of the medium from the outside. The value of $\alpha = 0$ for the radiation propagated downwards ($\vartheta \leq \pi/2$) and $\alpha = z_0 \sec\vartheta$ (z_0 is the geometrical thickness of the layer) for the radiation propagated upwards.

This integral can be found analytically. The solution is:

$$Y_i(x) = (H^{-1})_{ij}\sigma^d_{sca\ jk}[a_{kl}P_{1i} + b_{kl}P_{2i} + c_{kl}P_{3i} + d_{kl}P_{4i}]J_l,$$

where
$$P_{mi}(x) = \frac{\exp(-\Lambda_i x)}{\Lambda_i - v\Lambda_m}[\exp(x(\Lambda_i - v\Lambda_m)) - \exp(\alpha(\Lambda_i - v\Lambda_m))].$$

Finally, the Stokes vector is given by the following equation: $\vec{I} = \hat{H}\vec{Y}$.
Note, it follows on the boundaries of a turbid layer:
$$P_{mi} = \frac{1 - \exp((\Lambda_i - \Lambda_m v)z_0 \sec \vartheta)}{\Lambda_i - \Lambda_m v}$$

for reflected light ($x = 0$ and $\alpha = z_0 \sec \vartheta$, where z_0 is the geometrical thickness of a layer) and
$$P_{mi} = \frac{\exp(-\Lambda_m z_0 \sec \vartheta_0) - \exp(-\Lambda_i z_0 \sec \vartheta)}{\Lambda_i - u\Lambda_m}$$

for transmitted light ($x = z_0 \sec \vartheta, \alpha = 0$).

Appendix 4

Local optical characteristics of cloudy media

The following parameterizations of the extinction coefficient σ_{ext}, the asymmetry parameter g, and the absorption coefficient σ_{abs} of cloudy media allow Mie calculations in visible and near-infrared to be avoided:

$$\sigma_{ext} = \frac{3C_v}{2a_{ef}}\left(1 + \frac{1.1}{(x_{ef})^{2/3}}\right), \tag{A4.1}$$

$$\sigma_{abs} = \frac{4\pi\chi C_v}{\lambda}\sum_{n=0}^{4} p_n(x_{ef})^n, \tag{A4.2}$$

$$1 - g = \sum_{n=0}^{4} q_n(x_{ef})^{-2n/3} \tag{A4.3}$$

Here C_v is the volumetric concentration of droplets, $x_{ef} = 2\pi a_{ef}/\lambda$, and χ is the imaginary part of the refractive index of the particles. The parameters p_n and q_n depend on the refractive index. They are given in Tables A4.1 and A4.2 for selected

Table A4.1. Parameters p_n at the wavelength 1.55 μm.

p_0	p_1	p_2	p_3	p_4
1.671	0.0025	$-2.365 \cdot 10^{-4}$	$2.861 \cdot 10^{-6}$	$-1.05 \cdot 10^{-8}$

Table A4.2. Parameters q_n for different wavelengths λ.

λ (μm)	q_0	q_1	q_2	q_3	q_4
0.65	0.1121	0.5118	0.8997	0.0	0.0
1.55	0.0608	2.465	-32.98	248.94	-636

wavelengths. The accuracy of Equations A4.1–A4.3 is better than 3% for the effective radius of droplets in the range 4–16 μm, which is most common for water clouds.

Clearly, the ratio of optical thicknesses for selected wavelengths is given by ratios of the values $(1 + 1.1/(x_{ef})^{2/3})$ for both wavelengths. Equations A4.1–A4.3 can be used to calculate the single scattering albedo, mean photon free path length and some other cloud local optical characteristics.

Appendix 5

Fresnel equations

The reflection of light from a semi-infinite homogeneous medium having the relative refractive index m can be found using so-called Fresnel equations. In particular, for the electric vector of a reflected electromagnetic wave (Born and Wolf, 1999):

$$E_l = R_l E_{0l},$$

where

$$R_l = \frac{\tan(\alpha)}{\tan(\beta)}$$

and $\alpha = \varphi - \psi$, $\beta = \varphi + \psi$, and $\psi = \arcsin(\sin(\varphi)/m)$. Here φ is the angle of incidence, measured from the normal to the reflecting boundary. The index l means that both the incident (E_{0l}) and reflected (E_l) waves are in the plane parallel to the incidence plane, which holds the reflected and incident waves and the normal.

If the electric vector of the incident wave is perpendicular to the plane of incidence, then we have (Born and Wolf, 1999):

$$E_r = R_r E_{0r},$$

where

$$R_r = -\frac{\sin(\alpha)}{\sin(\beta)}.$$

Clearly, it follows for the ratio of energies ($\sim E^2$) of reflected and incident light that:

$$\rho = \tfrac{1}{2}[|R_l|^2 + |R_r|^2],$$

where we have assumed that the incident light is unpolarized. We can define the degree of linear polarization by:

$$p_l = \frac{|R_r|^2 - |R_l|^2}{|R_r|^2 + |R_l|^2},$$

which is positive for oscillations predominantly perpendicular to the planes of incidence. Note that it follows that the reflectivity $\rho \leq 1$. For non-absorbing media the difference $t = 1 - \rho$ is the transmissivity. We also have:

$$t = \tfrac{1}{2}[t_l + t_r],$$

where

$$t_l = 1 - |R_l|^2, \qquad t_r = 1 - |R_r|^2.$$

We can also define the degree of polarization of transmitted light as:

$$p_l = \frac{t_r - t_l}{t_r + t_l}.$$

If a medium has a finite thickness h, then the reflectivity $\Re$ is:

$$\Re = \rho + \frac{(1-\rho)^2 \rho \sigma^2}{1 - \rho^2 \sigma^2}.$$

Here the effects of the interference are neglected and the value of $\sigma = \exp(-(4\pi\chi/\lambda)h\sec\psi)$ accounts for absorption along the path $h\sec\psi$. Here ψ is the refraction angle and χ is the imaginary part of the refractive index. This result was first noted by Stokes (1862). He also obtained for the transmissivity T in this case:

$$T = \frac{(1-\rho)^2 \sigma}{1 - \rho^2 \sigma^2}.$$

The difference $A = 1 - \Re - T$ is called the absorptivity. It is given by

$$A = \frac{(1-\rho)(1-\sigma)}{1 - \rho\sigma}.$$

We present the degree of polarization p_l for the reflected and transmitted light at $m = 1.13$, 1.33, and 1.5 in Figures A5.1 and A5.2.

The degree of polarization of reflected light has a maximum at the incidence angle $\tan(m)$, which is called the Brewster angle. At this angle reflected light is polarized perpendicular to the plane of incidence. The degree of polarization of reflected light is equal to zero at the angles of incidence 0 and 90°. The maximum of the degree of polarization of the reflected light moves to smaller angles of incidence with decreasing refractive index.

The degree of polarization of the transmitted light is generally smaller than that of the reflected light. It increases with increasing refractive index. Oscillations occur predominantly in the plane parallel to the incidence plane.

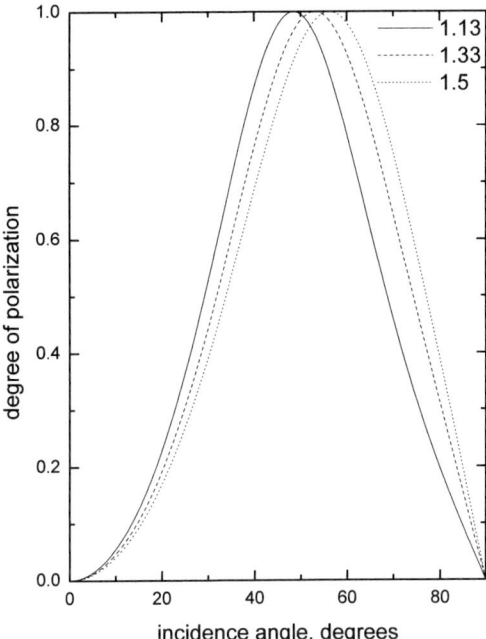

Figure A5.1. The dependence of the degree of polarization of the reflected light on the incidence angle for selected values of the refractive index $n = 1.13$, 1.33, and 1.5.

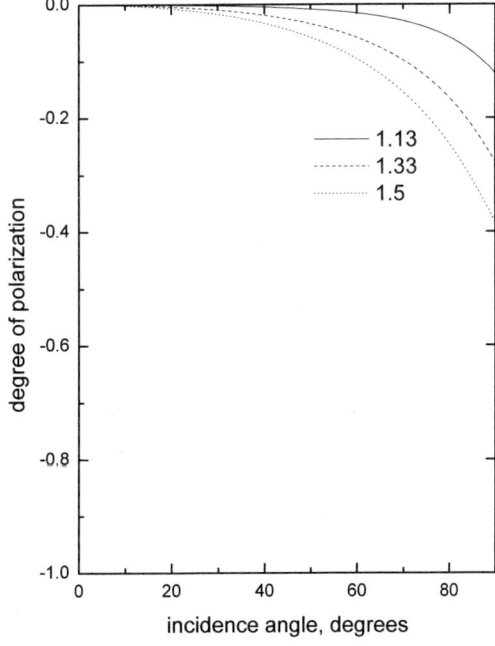

Figure A5.2. The same as in Figure A5.1 but for the transmitted light.

FURTHER READING

Born, M. and Wolf, E. (1999) *Principles of Optics*. Cambridge University Press, Cambridge.

Brewster, D. (1815) On the laws which regulate the polarization of light by reflexion from transparent bodies. *Phil. Trans.*, **105**, 125–130, 158–159.

Fresnel, A. (1866) Memoire sur la reflexion de la lumière polarisée. In *Oeuvres Complètes de Fresnel*, **1**, 767—775.

Newton, I. (1730) *Optics*. 4th Edition, London, 354–361.

Stokes, G. (1862) On the intensity of the light reflected from or transmitted through a pile of plates. *Proc. Royal Soc.*, 23 January, 145–156.

Swindel, W. (ed) (1975) *Polarized Light: Benchmark Papers in Optics*. Halsted Press, New York.

Index

absorption
   coefficients 13–18, 211
   cross sections 14, 88, 90–1, 127, 128–9
   Fresnel equations 214
   matrix 45
   spherical polydispersions 68
ADT *see* anomalous diffraction theory
aerosols
   humidity 131
   optical thickness 94–5
   polarimetry 159–171, 181, 184
   remote sensing 94–5, 181, 184
   volumetric concentration 166
air bubbles 69–86
Airy theory 62–3
albedo
   scattering 18, 27–8, 181, 212
   spherical 28, 30–1
amplitude coefficients 126
amplitude scattering functions 186
amplitude scattering matrices
   extinction matrices 44
   optically active particles 133
   phase matrices 38–42
   radiative transfer 20
   Rayleigh particles 146–8
   Rayleigh–Gans approximation 128
   spherical particles 47
angles
   see also scattering . . .

azimuth 86, 134, 142, 148
Brewster 64, 214
correlation 124
ellipticity 7–10, 134, 142, 148
Euler 86
maximal refraction 64
observation 28–32
rainbow 153–4
refraction 64, 214
solar 168–9, 170–1
solid 49
zenith 86, 153–4
angular dependency
   clouds 154–5, 158
   ellipsoidal particles 95
   irregularly shaped particles 116–17
   optically active particles 137–41
angular functions 47
angular variation 155
anisotropic media 2, 7, 40, 45–6
anomalous diffraction theory (ADT) 185
approximation relative errors 177, 179
aspect ratio 103, 105–9
asymmetrical media 23, 45–6
asymmetry parameters
   bubbles 79, 84–5
   cloudy media 211
   diffused light 27–8
   droplets 79, 84–5

asymmetry parameters (*cont.*)
  Koch fractals 111, 115
  Mie theory 47–8
asymptotic regime vector 26–7
asymptotic theory 181, 184
atmospheric air 167–8, 170
atmospheric phase functions 94–5
average radii 124
averaged intensity 26–7
averaged vector matrices 25–6, 29–30
axis ratio 94, 98, 101–4
azimuth
  angles 86, 134, 142, 148
  averaged intensity 26–7
  averaged vector matrices 25–6
  introduction 7–8

bare soil 171, 174
Bessel functions, modified spherical 124
Bessel and Hankel functions 47, 133
bidirectional polarized reflectance 179–180
bio-particles 132
Bouguer law 201
Brewster angles 64, 214
broad minima 116
bubbles 75–86

CD *see* circular dichroism
chiral media 40, 132–6, 142–6
chiro-optical spectroscopy (CS) 184–92
circular cylinders 5, 108
circular dichroism (CD) 136, 141–6, 185–8, 192
circular polarization
  bubbles 77, 79, 85–6
  Jones matrices 203–4
  Mueller matrices 203–4
  polydispersions 49–61, 65–6
  radiative transfer equation 199
  Stokes vectors 205
  tensors 199
cirrus clouds 157–9
classification
  aerosols 159
  disperse media 45–6
clouds 28–32, 153–9, 181–4
cloudy media 211–12
coated ellipsoids 127–8

coated spheres 129–31
coefficients
  absorption 13–18, 211
  amplitude 126
  extinction 13–18, 211
  scattering 13–18
  variance 3, 50–61, 67, 124
coherence 18–19, 42–5, 200
Coleus blumei 172, 174
complex electric vectors 6
complex numbers 186–8
concentric spheres 126–31
conservation of energy 199
Continental aerosols 166
correlation angles 124
covariant radiative transfer equation (CRTE) 201
cross sections
  absorption 14, 88, 90–1, 127, 128–9
  differential scattering 15–16
  extinction 14
  scattering 14–16, 88, 90–1, 127
CRTE *see* covariant radiative transfer equation
crystalline clouds *see* ice ...
cubes 5, 110
cylinders 5, 104–10, 188–9

degree of polarization
  aerosols 159–171
  bubbles 69–79, 81, 85–6
  clouds 153–4, 182–4
  cylindrical particles 103
  diffused light 29–33
  ellipsoidal particles 95
  Fresnel equations 214–15
  ice clouds 157–9
  introduction 7–9
  Koch fractals 116–17
  land 171–3
  oceanic water 176–7
  optically active particles 134
  polydispersions 49–61, 63–6, 73
  radiative transfer equation 198
  Rayleigh ellipsoids 89–91
  skylight spectra 168–170
  spherical polydispersions 49–61, 63–6, 73
  tensors 198
density matrices 6–7, 10–12, 201

depolarizer matrices 204
diagonalized matrices 21–2
diagonalized tensors 88
dichroism 44–5, 136, 141–6, 185–8, 192
differential scattering cross section 15–16
diffraction 54–6, 185
diffused light 18–19, 20–33, 200, 201
dimensionless phase matrix 132–3
dipoles 87–8, 90–1
direct light 18–20
disperse media 1–12, 45–6
droplets 69–86, 181–2
dust aerosols 117, 121–3, 159–68, 170
dyadic tensor notation 198

Earth's reflectance 180
effective radii 3, 5, 50–61
effective refractive index 43
effective solid angles 49
effective variance 3–4
eigenvalues 22–3
eigenvectors 21–2
electric fields 90–1
electric vectors 6, 10–11, 213
electromagnetism 1–4, 10
ellipsoids 86–104
　chiro-optical spectroscopy 189–92
　polarizability tensors 92, 127–8
ellipticity
　angles 7–10, 134, 142, 148
　introduction 7–10
energy conservation law 199
energy ratio 213–14
entropy
　aerosols 162, 164–7
　bubbles 75, 78
　polydispersions 50, 53, 56, 61, 67, 69
environmental polarimetry 153–92
　aerosols 159–70
　chiro-optical spectroscopy 184–92
　clouds 153–9
　ice 179–80
　natural surfaces 170–80
　remote sensing 180–4
　snow 179–80
　turbid media 184–92
errors 178–9, 182–3
escape functions 27–9
Euler angles 86

experimental results, irregularly shaped particles 118–22
extinction
　coefficient 13–18, 211
　cross section 14
　efficiency 47–8
　matrices 21–3, 42–5, 136, 142–8
　process 199

fixed azimuth 27–8
flux density 24–5
forward–backward asymmetry 79, 84–5
fractal particle model (FPM) 118–19, 120, 124–5
fractals 110–19, 120, 124–5
Fraunhofer diffraction 110–11
free parameters 179
Fresnel equations 170, 213–15

Gamma functions 3–4
gamma particle size distribution 3–4, 50
Gaussian particle model (GPM) 122–5
general equations, local optical characteristics 37–46
generalized Bouguer law 201
geometrical factors, tensors 92–3, 128
geometrical scattering regions 54–6
global transmittance 30, 33
glory scattering 62–4, 66, 157
GPM see Gaussian particle model
gyrotropic tensors 2

half-wave plates 204
half-width 67
halo phenomena 104–10
Hankel see Bessel and Hankel functions
hexagonal cylinders 5, 105–10
homogenous particles 37–126
horizontal polarization 205
humidity 131

ice
　clouds 118–19, 157–9
　polarimetry 179–80, 184
imaginary part, refractive index 67–8
incident light flux density 24–5
incoherence 20–33, 199
induced dipole moments 90–1

industrial aerosols 166
infrared spectrum 155
inhomogeneous particles 126–31
intensity
  averaged 26–7
  diffused light 21–2
  distribution 153–9
  introduction 8
  invariant 198
  vector radiative transfer equation 13
interaction matrices 10–12, 38, 199, 202, 203–4
interface reflection matrices 174
internal structure 2
invariants 198
inversion algorithm 182–3
irregularly shaped particles 110–25
isotropic media 45–6, 47–8

Jones matrices 10–12, 38, 199, 202, 203–4
Junge particle size distribution 129–31

Koch fractals 110–18
Kronecker symbol 11

land, polarimetry 171–4
leaves 184
Legendre polynomials 133
Levi–Cevita tensors 198
light beams 4–10, 197–8, 201
linear differential equations 207–9
linear interaction 10
linear polarization
  bubbles 76, 78–9, 85–6
  diffused light 29
  Fresnel equations 214–15
  ice clouds 157–8
  Jones matrices 203
  Mueller matrices 203
  optically active particles 134–5
  polydispersions 50–61, 63–4, 66
  Stokes vectors 205
liquid water paths 183–4
loam samples 171, 174
local optical characteristics 37–148
  cloudy media 211–12
  cylindrical particles 103–10
  ellipsoidal particles 86–104

  general equations 37–46
  homogenous particles 37–126
  inhomogeneous particles 126–31
  irregularly shaped particles 110–25
  optically active particles 132–48
  spherical particles 47–86
log-normal particle size distribution 3–4

maritime aerosols 166
matrices
  see also extinction ...; normalized phase ...; phase ...; scattering ...
  averaged vector 25–6, 29–30
  density 6–7, 10–12, 201
  depolarizer 204
  dimensionless phase 132–3
  interaction 10–12, 38, 199, 202, 203–4
  interface reflection 174
  Jones 10–12, 38, 199, 202, 203–4
  Mueller 11–12, 19–20, 38–42, 203–4
  reflection ... 28, 174
  Stokes 18
  symbol 40
  unity 19–20
maximal refraction angle 64
Maxwell's equations 38
mean free path lengths 212
mean radius 3
Mie theory 47–8
mineral surfaces 171–4
models
  fractal particle 118–19, 120, 124–5
  Gaussian particle 122–5
  stochastic particle 110
modified spherical Bessel functions 124
Mojave Desert sand 171–4
molecular asymmetries 186–7
moments, induced dipole 90–1
Mueller matrices 11–12, 19–20, 38–42, 203–4
multiple light scattering 17–18, 153–5
multipliers 187–9, 191

nadir illumination 32–3
natural surfaces 170–80
needles 93
noctilucent clouds 159
non-spherical particles 4, 186–8
normalized phase matrices 104–9

asymmetrical media 45–6
dust aerosols 121, 122
ellipsoidal particles 94–104
isotropic media 45–6
Koch fractals 112–14, 115–16, 229
oceanic water 120, 175
optically active particles 135–6, 137–40
polydispersions 48–9, 56, 63–4
radiative transfer 18
Rayleigh–Gans approximation 128–9
spherical particles 45–6
spherical polydispersions 48–9, 56, 63–4
number concentration 2

oblate spheroids 5, 93, 94, 103, 104
observation angles 28–32
oceanic aerosols 159–69
oceanic water 120–1, 170–2, 174–80
operators 200
optical depth
   aerosols 167–5, 168–71
   bubbles 84–5
   clouds 154–6
   cloudy media 212
   degree of polarization 33
   radiative transfer 18
   vector radiative transfer equation 15
   wavelength dependence 181–3
optical properties 48–69, 70–4
   see also local optical characteristics
optical rotatory dispersion (ORD) 136, 141, 143, 145–6, 185–8, 192
optically active particles 22–5, 132–48
optically thick turbid media 26–7
ORD *see* optical rotatory dispersion
oscillating dipoles 87–8

particle size distributions (PSD)
   aerosols 159
   clouds 183–4
   coated spheres 129–31
   disperse media 2–3
   gamma 3–4, 50
   half-width influence 67
   spherical polydispersions 50–61, 67
particles
   see also ellipsoids; spherical particles
   bio-particles 132

cylinders 5, 104–10, 188–9
electromagnetism 1–4
fractals 118–19, 120, 124–5
gamma distribution 3–4, 50
homogenous 37–126
inhomogeneous 126–31
irregularly shaped 110–25
Junge distribution 129–31
log-normal distribution 3–4
non-spherical 4, 186–8
optically active 22–5, 132–48
orientation 2
stochastic model 110
PGA *see* poly-L-glutamic acid
phase functions
   aerosols 159–69
   bubbles 75–86
   coated spheres 131
   cylindrical particles 103–9
   dust aerosols 122–3
   ellipsoidal particles 94–101
   fractals 111–14, 118, 124–5
   Gaussian particle model 124–5
   ice clouds 118, 157
   Koch fractals 111–14, 118
   Mie theory 47–8
   optically active particles 136
   polydispersions 51–62, 71–4
   Rayleigh ellipsoids 88–91
   Rayleigh–Gans approximation 128–9
   vector radiative transfer equation 16
phase matrices
   aerosols 159–65, 167–9
   bubbles 75–86
   dust aerosols 121–3
   general equations 37–46
   Koch fractals 111
   oceanic water 174–5, 177–8
   optically active particles 134
   radiative transfer 18
   Rayleigh ellipsoids 88–91
   Rayleigh particles 147–8
phase plates 204
photons 37, 199
plane-parallel media radiative transfer 16
plants 184
plates 93
polar mesospheric clouds 159
polarimetry 153–92

polarizability tensors 87–8, 92, 127–8
polarization curve transformation 31–2
Polarization and Directionality of Earth Reflectances (POLDER) 180
polarization ellipse characteristics 9
polarization enhancement factor 33
polarized radiative transfer 13–33
POLDER *see* Polarization and Directionality of Earth Reflectances
poly-L-glutamic acid (PGA) 136–41, 143–5, 192
polydispersions 48–69, 70–4, 129–31
prolate spheres 104, 106–7
prolate spheroids 5, 93, 94
propagation 17–18
PSD *see* particle size distributions

quarter-wave plates 204
quartz aerosol particles 121–3

radiative transfer 13–33
  diffused light 20–33
  direct light 18–20
  equations 13–18, 172, 173, 197–202
  Rayleigh ellipsoids 90
  vector equations 13–18, 181, 197
radii
  average 124
  effective 3, 5, 50–61
  Gaussian particle model 124
  mean 3
rainbow
  angles 153–4
  scattering 62–6, 70, 101, 157, 184
ratios
  aspect 103, 105–9
  axis 94, 98, 101–4
  energy 213–14
Rayleigh ...
  approximation 127–8, 146–8
  Gans approximation 128–9
  particles 30, 50, 87–93
  scattering 63, 67, 135–6
real part, refractive index 68–9
reflectance spectra 170–2, 174
reflection ...
  bidirectional polarized 177, 180

degree of polarization 214–15
functions 24–6, 95, 170–2, 174, 183–4
intensity 153–5, 180–1
linear differential equations 209
matrices 28, 172, 173
specular 179
vector functions 24–6
refraction angles 64, 214
refractive index
  bubbles 69, 79, 84–5
  clouds 155
  cloudy media 211
  effective 43
  Fresnel equations 214
  imaginary part 67–8
  introduction 1–2
  optically active particles 142–4
  polydispersions 67–70
  real part 68–9
  relative 86
relative effective solid angles 49
relative refractive index 86
remote sensing 94–5, 171, 180–4
rotators 204

sand 171, 173
satellite aerosol remote sensing 94–5
scalar radiative transfer equation 201
scattering
  see also phase functions
  albedo 18, 27–8, 181, 212
  chiral media 132
  coefficient 13–18
  cross sections 14, 88, 90–1, 127
  efficiency 47–8
  global transmittance 28
  isotropic homogenous spheres 47–8
scattering angles
  bubbles 71–80, 82–3
  coated spheres 131
  cubes 110
  ellipsoidal particles 95–100
  hexagonal particles 105–10
  ice clouds 157–9
  Koch fractals 112–14, 116–18
  oceanic water 175–6
  optically active particles 137–41
  polydispersions 51–61, 68–9

Rayleigh ellipsoids 89–90
  spherical polydispersions 51–61, 68–9
  vector radiative transfer equation 15
scattering matrices
  diffused light 22–3
  optically active particles 132–4
  phase matrices 38–42
  radiative transfer 18
  Rayleigh particles 146–8
  total 45
SCATTERLIB 38
semi-axes 86
shape characteristics 4
single scattering albedo 27–8, 181, 212
size parameters 2, 4, 86
  see also particles
skylight polarization spectra 168–71
slabs 25–33
snow 177, 180
soft Rayleigh ellipsoidal particles 90–1
solar angles 168–71
solar elevation 181
solid angles 49
soot aerosols 159–69
specific dependency 190–1
spectra 142, 168–71
spectroscopy 184–92
specular reflection 177
spheres
  axis ratio 103, 104
  cylindrical particles 104, 106–7
  disperse media 5
  geometrical factors 93
spherical albedo 30–1
spherical particles 47–86
  bubbles 75–86
  Mie theory 47–8
  normalized phase matrices 45–6
  polydispersion 48–69, 70–4
spheroids 5, 93–108
  see also ellipsoids
stochastic particle model 110
Stokes matrix 18
Stokes parameters 5–9, 39–40, 49
Stokes vectors
  diffused light 19, 21–4
  direct light 18–19
  extinction matrices 43
  general polarization 205

interaction matrices 10–12
introduction 5–9
linear differential equations 209
optically active particles 134
phase matrices 38–40
polarimetry 180
polydispersions 49
radiative transfer equation 17–18, 197–8
remote sensing 180
tensors 197–8
supernumerary bows 63–6
surface areas 5
surface reflectance 170–2, 174
symbol matrix 40

tensors
  introduction 2, 6–7
  light beams 6–7, 197–8, 201
  polarizability 87–8, 92, 127–8
  radiative transfer equation 197–202
tetrahedrons 5
thick layer diffused light 25–33
thin layer diffused light 20–5
total scattering matrix 45
traces 11, 22, 23
transfer equations 13–18, 174, 181, 197–202
transformation law 11
transmission
  clouds 183
  degree of polarization 214–15
  Fresnel equations 214
  linear differential equations 209
  vector functions 25–6
transpose tensors 199
turbid media 184–92, 209
two-layered spheres 126–31

uniform spherical Rayleigh particles 91–2
unity matrices 19–20
unity vectors 27
urban aerosols 166

variance
  coefficient 3, 50–61, 67, 124
  effective 3–4
vectors
  see also Stokes ...

vectors (*cont.*)
  asymptotic regime 26–7
  asymptotic theory 181
  averaged matrices 25–6, 29–30
  eigenvectors 21–2
  electric vectors 6, 10–11, 213
  functions 24–6
  radiative transfer equation 13–18, 181, 197
  unity 27
  wave 38
vegetation 170–1, 174
vertical polarization 205
volume 5
volumetric concentration 14, 166, 177, 180

water
  clouds 153–6
  droplets 69–86
  oceanic 120–1, 169, 172, 174–80, 184
  soluble aerosols 159–69
wave numbers 20
wave propagation 207
wave vectors 38
wavelengths 20, 155–6, 181
World Metrological Organization 159, 166

zenith angles 86, 153–4
zenith direction 168–9, 170–1
zero polarization 65, 157, 167

Printing: Mercedes-Druck, Berlin
Binding: Stein+Lehmann, Berlin